American Book Company's

Mastering the Georgia Accelerated Mathematics I Course

Erica Day

Colleen Pintozzi

American Book Company

P. O. Box 2638

Woodstock, Georgia 30188-1383

Toll Free 1 (888) 264-5877 Phone (770) 928-2834

Toll Free Fax 1 (866) 827-3240

WEB SITE: www.americanbookcompany.com

Acknowledgements

In preparing this book, we would like to acknowledge Eric Field for his contributions in developing graphics for this book and Camille Woodhouse for her contributions in editing and formatting this book. We would also like to thank our many students whose needs and questions inspired us to write this text.

Contents

Contents

Contents

Contents

Preface

Mastering the Georgia Accelerated Mathematics I Course will help you review and learn important concepts and skills related to algebra, geometry, and data analysis. First, take the Diagnostic Test beginning on page 1 of the book. Next, complete the evaluation chart with your instructor in order to help you identify the chapters which require your careful attention. When you have finished your review of all of the material your teacher assigns, take the practice tests to evaluate your understanding of the material presented in this book. **The materials in this book are based on the Georgia Performance Standards that are published by the Georgia Department of Education. Also, the chapters are designed to follow the Curriculum Map published on the Georgia Department of Education website. The complete list of standards is located in the Answer Key. Each question in the Diagnostic and Practice Tests is referenced to the standard, as is the beginning of each chapter.**

This book contains several sections. These sections are as follows: 1) A Diagnostic Test; 2) Chapters that teach the concepts and skills for ***Mastering the Georgia Accelerated Mathematics I Course***; and 3) Two Practice Tests. Answers to the tests and exercises are in a separate manual.

Even though this book covers all accelerated mathematics I standards, the three tests only cover the mathematics I standards. Students taking an accelerated mathematics I course will learn both mathematics I and some mathematics II standards, but they will only be tested on mathematics I standards.

ABOUT THE AUTHORS

Erica Day has a Bachelor of Science Degree in Mathematics and is working on a Master of Science Degree in Mathematics. She graduated with high honors from Kennesaw State University in Kennesaw, Georgia. She has also tutored all levels of mathematics, ranging from high school algebra and geometry to university-level statistics, calculus, and linear algebra. She is currently writing and editing mathematics books for American Book Company, where she has coauthored numerous books, such as *Passing the Georgia Algebra I End of Course*, *Passing the Georgia High School Graduation Test in Mathematics*, *Passing the Arizona AIMS in Mathematics*, and *Passing the New Jersey HSPA in Mathematics*, to help students pass graduation and end of course exams.

Colleen Pintozzi has taught mathematics at the middle school, junior high, senior high, and adult level for 22 years. She holds a B.S. degree from Wright State University in Dayton, Ohio and has done graduate work at Wright State University, Duke University, and the University of North Carolina at Chapel Hill. She is the author of many mathematics books including such best-sellers as *Basics Made Easy: Mathematics Review*, *Passing the New Alabama Graduation Exam in Mathematics*, *Passing the Louisiana LEAP 21 GEE*, *Passing the Indiana ISTEP+ GQE in Mathematics*, *Passing the Minnesota Basic Standards Test in Mathematics*, and *Passing the Nevada High School Proficiency Exam in Mathematics*.

GA Accelerated Mathematics I Formula Sheet

Below are the formulas you may find useful as you work the problems. However, some of the formulas may not be used. You may refer to this page as you take the test.

Area

Rectangle and Parallelogram $A = bh$

Triangle $A = \frac{1}{2}bh$

Circle $A = \pi r^2$

Trapezoid $A = \frac{1}{2}(h)(b_1 + b_2)$

Circumference

$C = \pi d \qquad \pi \approx 3.14$

Volume

Rectangular Prism/Cylinder $V = Bh$

Pyramid/Cone $V = \frac{1}{3}Bh$

Surface Area

Rectangular Prism $SA = 2lw + 2wh + 2lh$

Cylinder $SA = 2\pi r^2 + 2\pi rh$

Pythagorean Theorem

$$a^2 + b^2 = c^2$$

Mean Absolute Deviation

$$\frac{\sum\limits_{i=1}^{N} \left| x_i - \overline{x} \right|}{N}$$

the average of the absolute deviations from the mean for a set of data

Expected Value

$$E(x) = \sum\limits_{i=1}^{n} x_i\, p(x_i)$$

the sum of each outcome multiplied by its probability of occurrence

Permutations

$$_nP_r = \frac{n!}{(n-r)!}$$

Combinations

$$_nC_r = \frac{n!}{r!(n-r)!}$$

Interquartile Range

the difference between the first quartile and third quartile of a set of data

Diagnostic Test

Part 1

1. Which is the graph of $f(x) = x^2$?

(A)

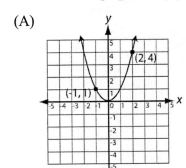

(B)

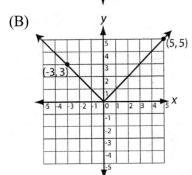

(C)

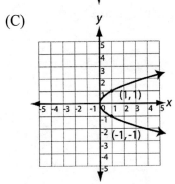

(D)

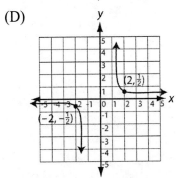

MA1A1b

2. $f(x) = 4x + 6$; What is $f(2)$?

(A) 10
(B) 12
(C) 14
(D) 16

MA1A1a

3. What is the domain of the function $f(x) = \dfrac{1}{x}$?

(A) $(-\infty, \infty)$
(B) $(-\infty, 0) \cup (0, \infty)$
(C) $(-\infty, 1) \cup (1, \infty)$
(D) $(-\infty, -1) \cup (-1, \infty)$

MA1A1d

4. What is the pattern in the following sequence $\{3, 4, 5, 6, 7, ...\}$?

(A) n
(B) $n + 1$
(C) $n + 2$
(D) n^2

MA1A1f

5. What is the intersection point of the graphs of the equations $f(x) = 4x - 1$ and $g(x) = x + 2$?

(A) $(1, 3)$
(B) $\left(\frac{1}{5}, 2\frac{1}{5}\right)$
(C) $(3, 1)$
(D) $(2, 4)$

MA1A1i

1

6. Subtract $y^2 + 4$ from $3y^2 + 8y$.

   (A) $4y^2 + 12y$
   (B) $4y^2 + 8y + 4$
   (C) $2y^2 + 8y - 4$
   (D) $2y^2 + 8y + 4$

   MA1A2c

7. Multiply $\dfrac{x^3 - x^2}{y^2}$ and $\dfrac{y^2}{x}$.

   (A) $\dfrac{x^4 - x^3}{y^2 x}$

   (B) $\dfrac{x^4 y^2 - 2x^3 y^2}{y^2 x}$

   (C) $\dfrac{x^3 y^2 - x^2 y^2}{y^2 x}$

   (D) $x^2 - x$

   MA1A2c

8. Factor the expression $x^2 + x - 30$.

   (A) $(x - 6)(x + 5)$
   (B) $(x - 10)(x + 3)$
   (C) $(x - 2)(x + 15)$
   (D) $(x + 6)(x - 5)$

   MA1A2e

9. Given the following measures of the angles of a triangle, find x. $90°, 70°, x$.

   (A) $10°$
   (B) $20°$
   (C) $30°$
   (D) $40°$

   MA1G3a

10. A square is _____ a rectangle.

   (A) Always
   (B) Sometimes
   (C) Never
   (D) Not enough information

   MA1G3d

11. Given the following triangle, which side is the longest?

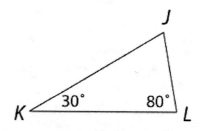

   (A) $\overline{KJ}$

   (B) $\overline{JL}$

   (C) $\overline{KL}$

   (D) Both $\overline{KJ}$ and $\overline{KL}$ are the same.

   MA1G3b

12. Given the two triangles, which theorem for triangles proves they are equivalent?

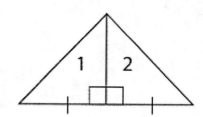

   (A) SSS

   (B) HL

   (C) SAS

   (D) ASA

   MA1G3c

13. A _____ creates 6 subtriangles, which have equal areas.

   (A) centroid

   (B) incenter

   (C) circumcenter

   (D) orthocenter

   MA1G3e

2   Copyright ©American Book Company

14. Sally is seven years old and goes to the store with her mother every Saturday morning. Every time they go, she notices that her mother gives the check out clerk money before they leave the store. Sally concluded that everyone has to give money in exchange for things they want to take out of the store. This an example of:

(A) deductive reasoning
(B) inductive reasoning
(C) analytical reasoning
(D) none of the above

MA1G2a

15. If I make all passing grades this week, then I will go to the movies on Saturday. What is the inverse of this statement?

(A) If I don't make all passing grades this week, then I won't go to the movies on Saturday.
(B) If I go to the movies on Saturday, then I made all passing grades this week.
(C) If I don't go to the movies on Saturday, then I didn't make all passing grades this week.
(D) If I don't make all passing grades this week, then I will go to the movies on Saturday.

MA1G2b

16. A = the number of math majors
B = the number present in class today
$A \cup B$ = math majors or present
$A \cap B$ = math majors and present
$n(A \cup B) = n(A) + n(B) - n(A \cap B)$
Using this information, find the probability that a student is a math major or present if $A = 36$, $B = 22$, and $A \cap B = 13$.

(A) 40
(B) 45
(C) 50
(D) 58

MA1D1a

17. How many permutations can you find using the letters in the word CAT?

(A) 3
(B) 6
(C) 9
(D) 12

MA1D1b

18. If Joe rolls 2 six-sided number cubes, the probability of him rolling a six and a double $(3, 3)$ is

(A) mutually exclusive.
(B) not mutually exclusive.
(C) both A & B.
(D) not enough information

MA1D2a

19. If Jonathan is rolling a six-sided number cube and spinning a spinner, these two events are

(A) independent.
(B) dependent.
(C) mutually exclusive.
(D) not mutually exclusive.

MA1D2b

20. Mrs. Smith was looking at her class' test scores. After two tests, 25% of her students had passed both tests. 42% passed the first test. What percent that passed the first test also passed the second?

(A) 17%
(B) 60%
(C) 67%
(D) 77%

MA1D2c

21. There are 24 candy-coated chocolate pieces in a bag. Eight have defects in the coating that can be seen only with close inspection. What is the probability of pulling out a defective piece without looking?

(A) 8

(B) $\dfrac{8}{23}$

(C) $\dfrac{1}{4}$

(D) $\dfrac{1}{3}$

MA1D2d

22. What is the mean absolute deviation of the following set of numbers $\{2, 7, 9, 1, 6, 5\}$?

(A) 1.5
(B) 2.8
(C) 2.33
(D) 3

MA1D4

23. The school cafeteria workers are conducting a survey of students' likes and dislikes for the foods they serve at lunch time. Once they tally the data and begin to order the food they need, which measure of central tendency will help most to make food purchases?

(A) Mean
(B) Mode
(C) Median
(D) Range

MA1D3a

24. $y = x^3 - 2$ is a(n) _____ function.

(A) odd
(B) even
(C) neither
(D) not enough information

MA1A1h

25. What are the solutions to $y = x^2 + 5x + 6$?

(A) $x = 3, 2$
(B) $x = -3, 2$
(C) $x = 3, -2$
(D) $x = -3, -2$

MA1A4b

26. $\sqrt{x} + 2 = 5$. What is x?

(A) 3
(B) $\sqrt{3}$
(C) 6
(D) 9

MA1A2b

27. According to the chart, what are the solution(s) (x-intercepts) to the equation $f(x) = x^2 + 2x + 1$?

x	$f(x)$
-2	1
-1	0
0	1
1	4
2	9

(A) -1

(B) 1

(C) $-1, 1$

(D) $1, -1$

MA1A4b

28. The points $(0, 0)$, $(5, 0)$, $(0, 4)$ form what type of triangle?

(A) Isosceles
(B) Equilateral
(C) Scalene
(D) Right

MA1G1e

29. Solve for x: $3(x - 2) - 1 = 6(x + 5)$

(A) -4

(B) $-\dfrac{37}{3}$

(C) 4

(D) $\dfrac{23}{3}$

MM1A3d

30. What is the distance between $(1, 1)$ and $(6, 6)$?

(A) $\sqrt{25}$

(B) 5

(C) $\sqrt{50}$

(D) 25

MA1G1a

31. What is the closest distance between $y = 3x - 2$ and $(0, 0)$? Use the following formula:

$$d = \frac{|am + bn + c|}{\sqrt{a^2 + b^2}}$$

(A) 0.63

(B) 0.67

(C) 0

(D) 1.23

MA1G1b

32. What is the midpoint of $(6, 4)$ and $(3, -4)$?

(A) $(4.5, 0)$

(B) $(0, 4.5)$

(C) $(-4.5, 0)$

(D) $(0, -4.5)$

MA1G1c

33. Samantha is a student at Etowah High School. Look at the following histograms. The first histogram displays the number of sodas consumed each day by each student in Samantha's math class. The second histogram displays the number of sodas consumed each day by each of the students in the ninth-grade class at her school.

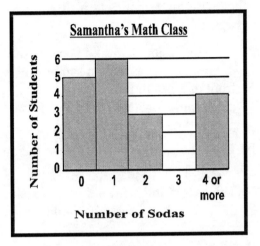

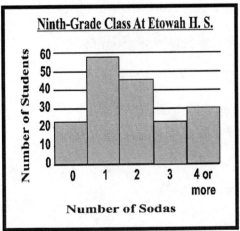

Find the difference in lower quartiles between the two data sets displayed in the histograms.

(A) 0

(B) 1

(C) 2

(D) 3

MA1D3b

34. Nicole is in charge of buying food for a political campaign picnic lunch in City Center Park. In order to determine which foods people would prefer, Nicole conducted a survey of food preferences in front of a small health-food store near the park. The results indicated that most people prefer fruits and vegetables. Is this conclusion valid? Use mathematics to justify your answer.

(A) The conclusion is invalid because she did not choose a random sample. In front of a health food store, there is a disproportionately high number of people who prefer health food.

(B) The conclusion is valid because she did choose a random sample.

(C) The conclusion is invalid because she did not choose a enough people for her sample.

(D) The conclusion is valid because she did survey people that are more knowledgeable about food.

MA1D3c

35. Use the points $A = (0,0)$, $B = (3,2)$, $C = (0,2)$ and the Pythagorean Theorem to find the distance between A and B. What is the distance?

(A) $\sqrt{13}$
(B) 13
(C) 5
(D) $\sqrt{5}$

MA1G1d

36. Two six-sided number cubes each have faces numbered 1 through 6. The two cubes are rolled at the same time. What is the probability that the sum of the two top faces will be either a 11 or a 5?

(A) $\dfrac{1}{9}$

(B) $\dfrac{1}{18}$

(C) $\dfrac{1}{6}$

(D) $\dfrac{1}{36}$

MA1D2a

Part 2

37. Which is the graph of $f(x) = \dfrac{1}{x}$?

(A)

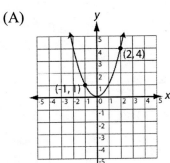

(B)

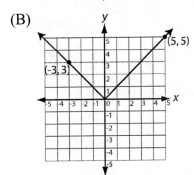

(C)

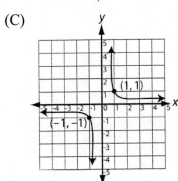

(D)

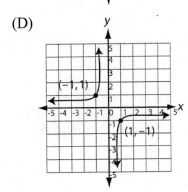

MA1A1b

38. What is the general shape of the function $y = x^2$?

(A) Straight line
(B) Parabola
(C) N shape
(D) V shape

MA1A1e

39. What is the rate of change of $f(x) = -\dfrac{3}{4}x + 2$?

(A) $-\dfrac{3}{4}$

(B) $\dfrac{3}{4}$

(C) 2

(D) -2

MA1A1g

40. What is $12\sqrt{3} - 3\sqrt{3}$?

(A) $4\sqrt{3}$
(B) $3\sqrt{3}$
(C) $9\sqrt{2}$
(D) $9\sqrt{3}$

MA1A2b

41. Multiply the binomials $(x + 2)(x + 1)$.

(A) $x^2 + 2x + 3$
(B) $x^2 + 2x + 2$
(C) $x^2 + 3x + 1$
(D) $x^2 + 3x + 2$

MA1A2c

42. What is the area of a triangle with height $h = 6 + x$ and base $b = 13 + x$?

(A) $x^2 + 19x + 78$

(B) $\dfrac{1}{2}x^2 + 19x + 39$

(C) $\dfrac{1}{2}x^2 + \dfrac{19}{2}x + 39$

(D) $\dfrac{1}{2}x^2 + \dfrac{19}{2}x + 78$

MA1A2f

43. What are the factors of $b^2 - 4b - 5$?

    (A) $(b - 5)(b + 1)$
    (B) $(b + 5)(b - 1)$
    (C) $(b - 3)(b + 2)$
    (D) $(b - 2)(b - 3)$

MA1A2e

44. Given the following triangle, find x.

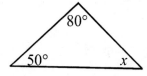

    (A) 30°
    (B) 40°
    (C) 50°
    (D) 60°

MA1G3a

45. Which of the following can be the measure of the third side of a triangle if the other two sides are 7 and 13?

    (A) 3
    (B) 4
    (C) 5
    (D) 8

MA1G3b

46. Given the two triangles, which theorem for triangles proves these are equivalent?

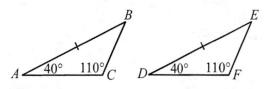

    (A) SSS
    (B) HL
    (C) SAS
    (D) AAS

MA1G3c

47. Given these two triangles, which theorem for triangles proves these are equivalent?

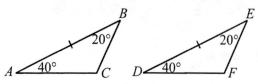

    (A) SSS
    (B) HL
    (C) SAS
    (D) ASA

MA1G3c

48. A rectangle is _____ a square.

    (A) always
    (B) sometimes
    (C) never
    (D) Not enough information

MA1G3d

49. The _____ is equidistant from the vertices of a triangle.

    (A) centroid
    (B) incenter
    (C) circumcenter
    (D) orthocenter

MA1G3e

50. A _____ is the center of an inscribed circle.

    (A) centroid
    (B) incenter
    (C) circumcenter
    (D) orthocenter

MA1G3e

51. All carnivores are meat eaters. Lions eat meat. Therefore, lions are carnivores. This kind of thinking is an example of _____ reasoning.

    (A) applied
    (B) inductive
    (C) qualitative
    (D) deductive

MA1G2a

52. If a figure is a triangle, then it has three sides. Which of the following statements is the contrapositive of the statement?

(A) If a figure is a triangle, then it has three sides.

(B) If a figure is not a triangle, then it does not have three sides.

(C) If a figure has three sides, then it is a triangle.

(D) If a figure does not have three sides, then it is not a triangle.

MA1G2b

53. At a toy manufacturing company in Nevada, the probability that they will make zero defective toys (out of every 100) is 0.27. The probability of making one defective toy is 0.44. What is the probability of making less than 1 defective toy?
(Hint: This is a mutually exclusive event.)

(A) 0.71

(B) 0.12

(C) 0.17

(D) 0.61

MA1D2a

54. If Jessica has a bag containing 3 blue bracelets and 7 green ones, what is the probability that she draws 1 blue then 1 green?

(A) $\frac{2}{9}$

(B) $\frac{15}{34}$

(C) $\frac{5}{9}$

(D) $\frac{7}{30}$

MA1D2b

55. If Sara has 6 shirts and 3 pairs of pants, how many different outfits does she have?

(A) 9

(B) 12

(C) 15

(D) 18

MA1D1b

56. Russell has a bag of black and red marbles. The probability that he first pulls out a black marble then pulls out a red marble is 34%. The probability he will pull out a black on the first try is 47%. What is the probability the second marble is red given the first marble is black?

(A) 53%

(B) 70%

(C) 72%

(D) 81%

MA1D2c

57. What is the probability that the spinner will stop on a shaded wedge or the number 1?

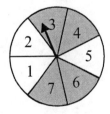

(A) $\frac{3}{7}$

(B) $\frac{5}{7}$

(C) $\frac{4}{7}$

(D) $\frac{6}{7}$

MA1D2d

58. Greg has test scores of 77, 90, 92, and 77 in English class. If he gets a 0 on his next paper, how will it affect the mean, median and mode of his scores?

(A) The mean will go down.
The median will go down.
The mode will remain the same.

(B) The mean will remain the same.
The median will remain the same.
The mode will remain the same.

(C) The mean, median and mode will all go down.

(D) The mean will go down.
The mode and median will remain the same.

MA1D3a

59. What is the mean absolute deviation of the following set of numbers $\{3, 6, 0, 9, 1, 7, 2\}$?

(A) 2
(B) 2.5
(C) 2.86
(D) 5

MA1D4

60. $2\sqrt{x} + 1 = 7$. What is x?

(A) 3
(B) 4
(C) 9
(D) 16

MA1A2b

61. What are the solutions to $y = x^2 - 7x + 12$?

(A) $x = -4, 3$
(B) $x = 4, 3$
(C) $x = -4, -3$
(D) $x = 4, -3$

MA1A4b

62. A Republican politician running for mayor wants to take a poll to determine if he is the favored candidate in the upcoming election. The mayoral candidate is scheduled to give a speech not at all related to the election at an upcoming town celebration. For convenience sake, his staff polls all of the people in attendance at the speech as to who they are planning on voting for in the upcoming mayoral election. The staff reports to the candidate that, based on the recent poll, he is favored by voters. Are the results of this poll most likely accurate or are they most likely misleading?

(A) The results of the survey are accurate as long as at least 100 people were polled.

(B) The results of the poll are most likely misleading because the sample of people surveyed were voluntarily attending a speech given by the candidate and do not necessarily represent the population of the town.

(C) The results of the poll are most likely accurate because people are generally honest in declaring who they are going to vote for in elections.

(D) The results of the poll are misleading because recent advertisements are encouraging people to vote, creating a more opinionated population than usual.

MA1D3c

63. What is the solution to the equations $f(x) = x + 3$ and $g(x) = -2x$?

(A) $(3, 6)$

(B) $(1, 4)$

(C) $(-1, 2)$

(D) $(-3, 6)$

MA1A1i

64. $f(x) = x^2 - x - 2$ According to the graph, what are the solutions (x-intercepts) to the equation?

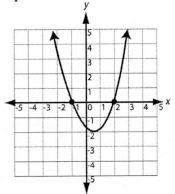

(A) $-2, 0$
(B) $-1, 1$
(C) $-1, 2$
(D) $-2, 1$

MA1A3c

65. What is the distance between $(0, 0)$ and $(3, 2)$?

(A) 13
(B) $\sqrt{13}$
(C) 4.12
(D) 3.555

MA1G1a

66. Solve for x: $x^2 - 3x - 10 = 0$

(A) $x = 5, 2$
(B) $x = 5, -2$
(C) $x = -5, 2$
(D) $x = -5, -2$

MA1A4b

67. What is the midpoint of $(-1, 2)$ and $(3, -6)$?

(A) $(1, 2)$
(B) $(2, 3)$
(C) $(3, 2)$
(D) $(1, -2)$

MA1G1c

68. Samantha is a student at Etowah High School. Look at the following histograms. The first histogram displays the number of sodas consumed each day by each student in Samantha's math class. The second histogram displays the number of sodas consumed each day by each of the students in the ninth-grade class at her school.

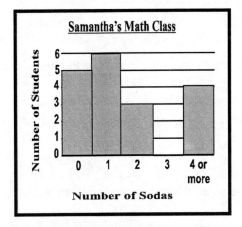

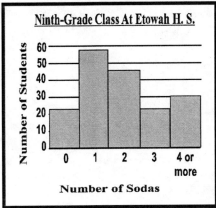

Find the difference in upper quartiles between the two data sets displayed in the histograms.

(A) 0

(B) 1

(C) 2

(D) 3

MA1D3b

69. Use the points $A = (1, -1)$, $B = (1, -5)$, $C = (4, -5)$ and the Pythagorean Theorem to find the distance between A and C. What is the distance?

(A) $\sqrt{5}$
(B) 2.3
(C) 5
(D) 25

MA1G1d

70. You are about to get a car for the first time. This means you will also get a licence plate. Each licence plate gets 3 numbers, then 3 letters. The numbers can be the digits from 0 to 9, and the letters can be any letter from A to Z. If there are no restrictions on what numbers or letters you can get, how many possible combinations of the licence plates are there?

(A) 108
(B) 17,576,000
(C) 760
(D) 12,245,000

MA1D1a

71. Graph the points $(2, 0)$, $(2, 2)$, $(0, 2)$, $(0, 0)$ to find out what shape they create. What shape do they create?

(A) a square
(B) a rectangle
(C) a triangle
(D) a rhombus

MA1G1e

72. Sean rolls a six-sided number cube with faces numbered 1 through 6, then tosses a coin. What is the probability that the coin has heads facing up, given that the number cube landed on 4?

(A) $\dfrac{1}{12}$

(B) $\dfrac{1}{3}$

(C) $\dfrac{1}{18}$

(D) $\dfrac{1}{2}$

MA1D2c

Evaluation Chart for the Diagnostic Mathematics Test

Directions: On the following chart, circle the question numbers that you answered incorrectly. Then turn to the appropriate topics (listed by chapters), read the explanations, and complete the exercises. Review the other chapters as needed. Finally, complete the *Mastering the Accelerated Georgia Mathematics I Course* Practice Tests to further review.

		Questions Part 1	Questions Part 2	Pages
Chapter 1:	Relations and Functions	2, 3, 4		14–30
Chapter 2:	Graphing Functions	1	37, 38, 39	31–53
Chapter 3:	Logic and Geometric Proofs	14, 15	51, 52	54–61
Chapter 4:	Coordinate Geometry	28, 30, 31, 32 35	65, 67, 69, 71	62–74
Chapter 5:	Triangles	9, 11, 12, 13	44, 45, 46, 47, 49, 50	75–87
Chapter 6:	Quadrilaterals and Other Polygons	10	48	88–95
Chapter 7:	Circles and Spheres			96–109
Chapter 8:	Permutations and Combinations	17	55, 70	110–121
Chapter 9:	Probability	16, 18, 19, 20 21, 36	53, 54, 56, 57, 72	122–141
Chapter 10:	Statistics	22, 23, 33, 34	58, 59, 62, 68	142–155
Chapter 11:	Complex Numbers			156–163
Chapter 12:	More Functions	5, 24	63	164–173
Chapter 13:	Roots	26	40, 60	174–181
Chapter 14:	Polynomials	6, 7	41, 42	182–199
Chapter 15:	Factoring	8	43	200–222
Chapter 16:	Solving Quadratic Equations and Inequalities	25, 27	61, 64	223–239
Chapter 17:	Quadratic Functions		66	240–255
Chapter 18:	Step and Piecewise Functions			256–273
Chapter 19:	Using Algebra for Data Analysis			274–302

Chapter 1
Relations and Functions

This chapter covers the following Georgia Performance Standards:

MA1A	Algebra	MA1A1a
		MA1A1d
		MA1A1e
		MA1A1f

1.1   Relations

A **relation** is a set of ordered pairs. The set of the first members of each ordered pair is called the **domain** of the relation. The set of the second members of each ordered pair is called the **range**.

Example 1:   State the domain and range of the following relation:

$$\{(2, 4), (3, 7), (4, 9), (6, 11)\}$$

Solution: Domain: $\{2, 3, 4, 6\}$ the first member of each ordered pair

Range: $\{4, 7, 9, 11\}$ the second member of each ordered pair

State the domain and range for each relation.

1. $\{(2, 5), (9, 12), (3, 8), (6, 7)\}$

2. $\{(12, 4), (3, 4), (7, 12), (26, 19)\}$

3. $\{(4, 3), (7, 14), (16, 34), (5, 11)\}$

4. $\{(2, 45), (33, 43), (98, 9), (43, 61), (67, 54)\}$

5. $\{(78, 14), (29, 67), (84, 49), (16, 18), (98, 46)\}$

6. $\{(-8, 16), (23, -7), (-4, -9), (16, -8), (-3, 6)\}$

7. $\{(-7, -4), (-3, 16), (-4, 17), (-6, -8), (-8, 12)\}$

8. $\{(-1, -2), (3, 6), (-7, 14), (-2, 8), (-6, 2)\}$

9. $\{(0, 9), (-8, 5), (3, 12), (-8, -3), (7, 18)\}$

10. $\{(58, 14), (44, 97), (74, 32), (6, 18), (63, 44)\}$

When given an equation with two variables, the domain is the set of x values that satisfies the equation. The range is the set of y values that satisfies the equation.

Example 2:   Find the range of the relation $3x = y + 2$ for the domain $\{-1, 0, 1, 2, 3\}$.
Solve the equation for each value of x given. The result, the y values, will be the range.

Given:			**Solution:**	
x	y		x	y
-1			-1	-5
0			0	-2
1			1	1
2			2	4
3			3	7

The range is $\{-5, -2, 1, 4, 7\}$.

Find the range of each relation for the given domain.

	Relation	**Domain**	**Range**
1.	$y = 5x$	$\{1, 2, 3, 4\}$	
2.	$y = |x|$	$\{-3, -2, -1, 0, 1\}$	
3.	$y = 3x + 2$	$\{0, 1, 3, 4\}$	
4.	$y = -|x|$	$\{-2, -1, 0, 1, 2\}$	
5.	$y = -2x + 1$	$\{0, 1, 3, 4\}$	
6.	$y = 10x - 2$	$\{-2, -1, 0, 1, 2\}$	
7.	$y = 3|x| + 1$	$\{-2, -1, 0, 1, 2\}$	
8.	$y - x = 0$	$\{1, 2, 3, 4\}$	
9.	$y - 2x = 0$	$\{1, 2, 3, 4\}$	
10.	$y = 3x - 1$	$\{0, 1, 3, 4\}$	
11.	$y = 4x + 2$	$\{0, 1, 3, 4\}$	
12.	$y = 2|x| - 1$	$\{-2, -1, 0, 1, 2\}$	

1.2   Functions

Some relations are also **functions**. A relation is a function if **for every element in the domain, there is exactly one element in the range**. In other words, for each value for x there is only one unique value for y.

Example 3: $\{(2,4),(2,5),(3,4)\}$ is **NOT** a function because in the first pair, 2 is paired with 4, and in the second pair, 2 is paired with 5. The 2 can be paired with only one number to be a function. In this example, the x value of 2 has more than one value for y: 4 and 5.

Example 4: $\{(1,2),(3,2),(5,6)\}$ **IS** a function. Each first number is paired with only one second number. The 2 is repeated as a second number, but the relation remains a function.

Determine whether the ordered pairs of numbers below represent a function. Write "F" if it is a function. Write "NF" if it is not a function.

1. $\{(-1,1),(-3,3),(0,0),(2,2)\}$ _____

2. $\{(-4,-3),(-2,-3),(-1,-3),(2,-3)\}$ _____

3. $\{(5,-1),(2,0),(2,2),(5,3)\}$ _____

4. $\{(-3,3),(0,2),(1,1),(2,0)\}$ _____

5. $\{(-2,-5),(-2,-1),(-2,1),(-2,3)\}$ _____

6. $\{(0,2),(1,1),(2,2),(4,3)\}$ _____

7. $\{(4,2),(3,3),(2,2),(0,3)\}$ _____

8. $\{(-1,-1),(-2,-2),(3,-1),(3,2)\}$ _____

9. $\{(2,-2),(0,-2),(-2,0),(1,-3)\}$ _____

10. $\{(2,1),(3,2),(4,3),(5,-1)\}$ _____

11. $\{(-1,0),(2,1),(2,4),(-2,2)\}$ _____

12. $\{(1,4),(2,3),(0,2),(0,4)\}$ _____

13. $\{(0,0),(1,0),(2,0),(3,0)\}$ _____

14. $\{(-5,-1),(-3,-2),(-4,-9),(-7,-3)\}$ _____

15. $\{(8,-3),(-4,4),(8,0),(6,2)\}$ _____

16. $\{(7,-1),(4,3),(8,2),(2,8)\}$ _____

17. $\{(4,-3),(2,0),(5,3),(4,1)\}$ _____

18. $\{(2,-6),(7,3),(-3,4),(2,-3)\}$ _____

19. $\{(1,1),(3,-2),(4,16),(1,-5)\}$ _____

20. $\{(5,7),(3,8),(5,3),(6,9)\}$ _____

1.3   Function Notation

Function notation is used to represent relations which are functions. Some commonly used letters to represent functions include f, g, h, F, G, and H.

Example 5: $f(x) = 2x - 1$; find $f(-3)$

   Step 1:   Find $f(-3)$ means to replace x with -3 in the relation $2x - 1$.
   $$f(-3) = 2(-3) - 1$$

   Step 2:   Solve $f(-3)$. $f(-3) = 2(-3) - 1 = -6 - 1 = -7$
   $$f(-3) = -7$$

Example 6: $g(x) = 4 - 2x^2$; find $g(2)$

   Step 1:   Replace x with 2 in the relation $4 - 2x^2$.
   $$g(2) = 4 - 2(2)^2$$

   Step 2:   Solve $g(2)$. $g(2) = 4 - 2(2)^2 = 4 - 2(4) = 4 - 8 = -4$
   $$g(2) = -4$$

Find the solutions for each of the following.

1. $F(x) = 2 + 3x^2$; find $F(3)$

2. $f(x) = 4x + 6$; find $f(-4)$

3. $H(x) = 6 - 2x^2$; find $H(-1)$

4. $g(x) = -3x + 7$; find $g(-3)$

5. $f(x) = -5 + 4x$; find $f(7)$

6. $G(x) = 4x^2 + 4$; find $G(0)$

7. $f(x) = 7 - 6x$; find $f(-4)$

8. $h(x) = 2x^2 + 10$; find $h(5)$

9. $F(x) = 7 - 5x$; find $F(2)$

10. $f(x) = -4x^2 + 5$; find $f(-2)$

1.4   Function Tables

Functions can also use a variable such as n to be the input of the function and $f(n)$, read "f of n," to represent the output of the function.

Example 7:   Function rule: $3n + 4$

n	$f(n)$
-1	1
0	4
1	7
2	10

Fill in the tables for each function rule below.

1. rule: $2(n-5)$

n	$f(n)$
0	
1	
2	
3	

4. rule: $2x(x-1)$

x	$f(x)$
1	
2	
3	
4	

7. rule: $n(n+2)$

n	$f(n)$
1	
2	
3	
4	

2. rule: $3x(x-4)$

x	$f(x)$
0	
1	
2	
3	

5. rule: $\dfrac{1}{n+3}$

n	$f(n)$
1	
2	
3	
4	

8. rule: $2x - 3$

x	$f(x)$
1	
2	
3	
4	

3. rule: $\dfrac{2-n}{2}$

n	$f(n)$
0	
2	
4	
6	
8	

6. rule: $4x - x$

x	$f(x)$
-2	
-1	
0	
1	
2	

9. rule: $3 - 2n$

n	$f(n)$
-2	
-1	
0	
1	
2	

1.5   Recognizing Functions

Recall that a relation is a function with only one y value for every x value. We can depict functions in many ways including through graphs.

Example 8:

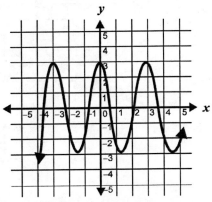

This graph **IS** a function because it has only one y value for each value of x.

Example 9:

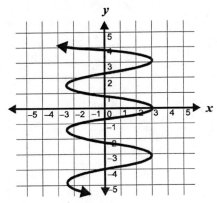

This graph is **NOT** a function because there is more than one y value for each value of x.

Hint: An easy way to determine a function from a graph is to do a vertical line test. First, draw a vertical line that crosses over the whole graph. If the line crosses the graph more than one time, then it is not a function. If it only crosses it once, it is a function. Take Example 9 above:

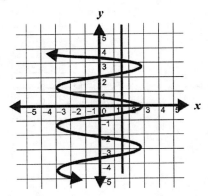

Since the vertical line passes over the graph six times, it is not a function.

Determine whether or not each of the following graphs is a function. If it is, write function on the line provided. If it is not a function, write NOT a function on the line provided.

1.

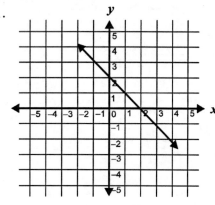

_____

2.

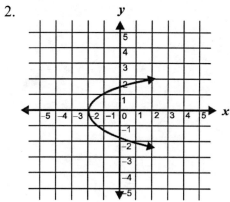

_____

3.

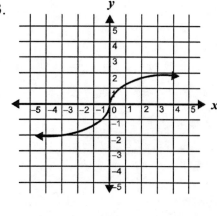

_____

4.

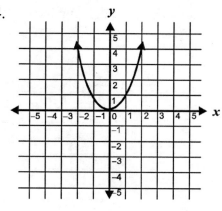

_____

5.

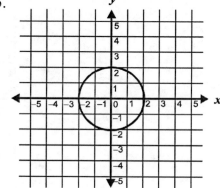

_____

6.

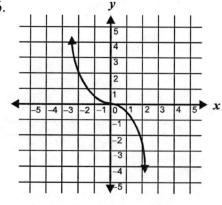

_____

7.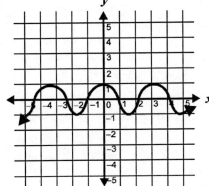

_____

8.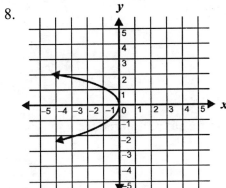

_____

9.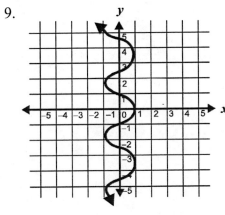

_____

10.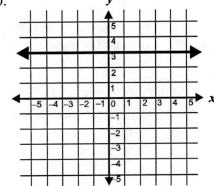

_____

11.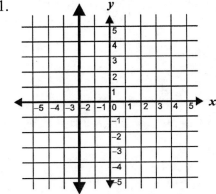

_____

12.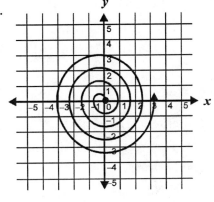

_____

1.6   Characteristics of Functions

There are many different types of functions. For our purposes, however, we will only look at the six basic functions and their characteristics.

The Linear Function: $f(x) = x$

Standard Form: $y = mx + b$, where m is the constant slope and b is the y-intercept

Characteristics: straight line

Domain: All Real Numbers

Range: All Real Numbers

Example 10: $f(x) = 2x + 7$; slope = 2, y-intercept = 7

x	-2	-1	0	1	2
$f(x)$	3	5	7	9	11

The Quadratic Function: $f(x) = x^2$

Standard Form: $f(x) = ax^2 + bx + c$, where $a \neq 0$, a, b, and c are real numbers, and c is the y-intercept

Characteristics: Has a vertex, parabola (U) shape

Domain: All Real Numbers

Range: $(0, \infty)$

Example 11: $f(x) = x^2 + 3x + 2$

x	-2	-1	0	1	2
$f(x)$	0	0	2	6	12

The Cubic Function: $f(x) = x^3$

Standard Form: $f(x) = ax^3 + bx^2 + cx + d$, where $a \neq 0$

Characteristics: N shape

Domain: All Real Numbers

Range: All Real Numbers

Example 12: $f(x) = x^3 + 3x^2 + 6x + 2$

x	-2	-1	0	1	2
$f(x)$	-6	-2	2	12	34

The Square Root Function: $f(x) = \sqrt{x}$

Characteristics: Never negative (neither x nor answer is negative), always contains a radical

Domain: $(0, \infty)$

Range: $(0, \infty)$

Example 13: $f(x) = \sqrt{x} + 2$

x	0	1	2	3	4
$f(x)$	2	3	$2 + \sqrt{2}$	$2 + \sqrt{3}$	4

The Absolute Value Function: $f(x) = |x|$

Characteristics: Answer usually positive unless outside value is being subtracted and is larger than inside value (i.e. $|-2| - 3 = -1$), V shape

Domain: All Real Numbers

Range: $(0, \infty)$

Example 14: $f(x) = |x| + 2$

x	-2	-1	0	1	2
$f(x)$	4	3	2	3	4

$f(x) = \dfrac{1}{x}$

Characteristics: Vertical asymptote at $x = 0$, undefined at $x = 0$

Domain: $(-\infty, 0) \cup (0, \infty)$

Range: $(-\infty, 0) \cup (0, \infty)$

Example 15: $f(x) = \dfrac{1}{x} + 2$

x	-2	-1	0	1	2
$f(x)$	$1\frac{1}{2}$	1	Undefined	3	$2\frac{1}{2}$

Determine the function described by the characteristics.

1. Its shape is a parabola.

2. It NEVER includes negative numbers.

3. It has an N shape.

4. It is undefined at $x = 0$.

5. The values of y are always positive, but x can be negative.

6. It is a straight line.

7. Of the 6 functions, how many have domains of all real numbers?

8. How many have limited ranges (not all real numbers)?

1.7   Sequences

In each of the examples below, there is a sequence of numbers that follows a pattern. Think of the sequence of numbers like the output for a function. You must find the pattern (or function) that holds true for each number in the sequence. Once you determine the pattern, you can find the next number in the sequence or any number in the sequence.

	Sequence	Pattern	Next Number	20th number in the sequence
Example 16:	$3, 4, 5, 6, 7$	$n + 2$	8	22

In number patterns, the sequence is the output (the range). The input (the domain) can be the set of whole numbers starting with 1. But, you must determine the "rule" or pattern. Look at the table below.

input	sequence
1 $\longrightarrow$	3
2 $\longrightarrow$	4
3 $\longrightarrow$	5
4 $\longrightarrow$	6
5 $\longrightarrow$	7

What pattern or "rule" can you come up with that gives you the first number in the sequence, 3, when you input 1? $n + 2$ will work because when $n = 1$, the first number in the sequence $= 3$. Does this pattern hold true for the rest of the numbers in the sequence? Yes, it does. When $n = 2$, the second number in the sequence $= 4$. When $n = 3$, the third number in the sequence $= 5$, and so on. Therefore, $n + 2$ is the pattern. Even without knowing the algebraic form of the pattern, you could figure out that 8 is the next pattern in the sequence. To find the 20th number in the pattern, use $n = 20$ to get 22.

	Sequence	Pattern	Next Number	20th number in the sequence
Example 17:	$1, 4, 9, 16, 25$	n^2	36	400
Example 18:	$-2, -4, -6, -8, -10$	$-2n$	-12	-40

Find the pattern and the next number in each of the sequences below.

	Sequence	Pattern	Next Number	20th number in the sequence
1.	$-2, -1, 0, 1, 2$			
2.	$5, 6, 7, 8, 9$			
3.	$3, 7, 11, 15, 19$			
4.	$-3, -6, -9, -12, -15$			
5.	$3, 5, 7, 9, 11$			
6.	$2, 4, 8, 16, 32$			
7.	$1, 8, 27, 64, 125$			
8.	$0, -1, -2, -3, -4$			
9.	$2, 5, 10, 17, 26$			
10.	$4, 6, 8, 10, 12$			

1.8 Using the $\sum$ Symbol

The Σ is used to write the sum of a series.

Series A $1 + 2 + 3 + \ldots + n + \ldots$

Series A starts with 1, +2, +3. Obviously the next item in the series would be 4.

Series B $1 + 4 + 7 + \ldots + (3n - 2) + \ldots$

Series B starts with 1, +4, +7 ...$(3n - 2)$. The expression in parenthesis shows how the pattern continues. The "n" in the expression is the item we're trying to find. We have 3 items: 1, 4, and 7. We are looking for the 4th item in the series so "n" = 4. $3n - 2$ where "n" = 4 is $3(4) - 2 = 10$. The next item in the series is 10.

Series B continues $1 + 4 + 7 + 10 + \ldots + (3n - 2) + \ldots$ to ∞ (infinity).

Series C $1 + 4 + 9 + \ldots + (n^2) + \ldots$

Replacing "n" with 4 in series C to find the 4th item in the series we have $4^2 = 16$. So now the series is $1 + 4 + 9 + 16 + \ldots + (n^2) + \ldots$

The sum of the first "n" terms of a series is called the **partial sum**, S_n. For example, in the series $3 + 6 + 9 + \ldots + 3n \ldots$

$$S_1 = 3 \qquad S_2 = 3 + 6 = 9 \qquad S_3 = 3 + 6 + 9 = 18$$

The Greek letter Σ, sigma, can be used to represent a sum. For example the third partial sum, S_3 of the series above can also be written as:

$$\sum_{k=1}^{3} 3k$$

So the series starts at $k = 1$ and ends at $k = 3$ and formula for the pattern is $3k$.

First $k = 1$ then 2, then 3. In expanded form we have: $3 + 6 + 9 = 18$. (We found the partial sum of the series.)

For each of the following write the Σ notation for each statement.

1. The sum of the series with the general term $k + 2$ from $k = 1$ to $k = 8$.

2. The sum of the series with the general term k^2 from $k = 1$ to $k = 5$.

3. The sum of the series with the general term $2k + 3$ from $k = 1$ to $k = 10$.

4. The sum of the series with the general term $2k - 6$ from $k = 3$ to $k = 8$.

5. The sum of the series with the general term $2m - 6$ from $m = 0$ to $m = 4$.

6. The sum of the series with the general term $2g^2$ from $g = 2$ to $g = 5$.

7. The sum of the series with the general term $g^2 + 2$ from $g = 3$ to $g = 7$.

8. The sum of the series with the general term $3h - 4$ from $h = 0$ to $h = 6$.

Example 19:   Write $\displaystyle\sum_{n=2}^{5} 3n$ in expanded form, then solve.

    Step 1:   In this problem, $n = 2, 3, 4$, and 5. Plug the values for n into the expression $3n$. We have $3 \cdot 2 + 3 \cdot 3 + 3 \cdot 4 + 3 \cdot 5$.

    Step 2:   Simplify the expanded form and solve.
$3 \cdot 2 + 3 \cdot 3 + 3 \cdot 4 + 3 \cdot 5 = 6 + 9 + 12 + 15 = 42$.

Write each of the following in expanded form and find the sums.

1. $\displaystyle\sum_{n=3}^{5} 2n$

2. $\displaystyle\sum_{n=0}^{4} n^2$

3. $\displaystyle\sum_{n=0}^{4} 2^n$

4. $\displaystyle\sum_{n=1}^{5} (2n - 1)$

5. $\displaystyle\sum_{n=0}^{3} (3n + 1)$

6. $\displaystyle\sum_{n=3}^{6} (3^n + 1)$

7. $\displaystyle\sum_{n=0}^{5} (2n + 1)$

8. $\displaystyle\sum_{n=1}^{4} (n - 6)$

9. $\displaystyle\sum_{n=4}^{6} (3n - 2)$

10. $\displaystyle\sum_{n=2}^{5} (6n - 4)$

Chapter 1 Review

1. What is the domain of the following relation? $\{(-1, 2), (2, 5), (4, 9), (6, 11)\}$

2. What is the range of the following relation? $\{(0, -2), (-1, -4), (-2, 6), (-3, -8)\}$

3. Find the range of the relation $y = 5x$ for the domain $\{0, 1, 2, 3, 4\}$.

4. Find the range of the relation $y = \dfrac{3(x-2)}{5}$ for the domain $\{-8, -3, 7, 12, 17\}$.

5. Find the range of the relation $y = 10 - 2x$ for the domain $\{-8, -4, 0, 4, 8\}$.

6. Find the range of the relation $y = \dfrac{4+x}{3}$ for the domain $\{-7, -1, 2, 5, 8\}$.

For each of the following relations given in questions 7–11, write "F" if it is a function and "NF" if it is not a function.

7. $\{(1, 2), (2, 2), (3, 2)\}$

8. $\{(-1, 0), (0, 1), (1, 2), (2, 3)\}$

9. $\{(2, 1), (2, 2), (2, 3)\}$

10. $\{(1, 7), (2, 5), (3, 6), (2, 4)\}$

11. $\{(0, -1), (-1, -2), (-2, -3), (-3, -4)\}$

For questions 12–17, find the range of the following functions for the given value of the domain.

12. For $g(x) = 2x^2 - 4x$; find $g(-1)$

13. For $h(x) = 3x(x - 4)$; find $h(3)$

14. For $f(n) = \dfrac{1}{n+3}$; find $f(4)$

15. For $G(n) = \dfrac{2-n}{2}$; find $G(8)$

16. For $H(x) = 2x(x - 1)$; find $H(4)$

17. For $f(x) = 7x^2 + 3x - 2$; find $f(2)$

Fill in the following function tables.

18. Rule: $\dfrac{(4-2n)}{2}$

n	$f(n)$
0	
1	
2	
3	
4	

19. Rule: $2n(n+1)$

n	$f(n)$
0	
1	
2	
3	
4	

20. Rule: $6n-3$

n	$f(n)$
2	
3	
4	
5	
6	

Find the pattern for the following number sequences, and then find the nth number requested.

21. 0, 1, 2, 3, 4   pattern_____

22. 0, 1, 2, 3, 4   20th number_____

23. 1, 3, 5, 7, 9   pattern_____

24. 1, 3, 5, 7, 9   25th number_____

25. 3, 6, 9, 12, 15   pattern_____

26. 3, 6, 9, 12, 15   30th number_____

Answer the following questions about the six basic functions.

27. What is the domain of a linear function?

28. What is the range of $f(x) = \dfrac{1}{x}$?

29. What function has a constant rate of change?

30. What is the standard form of a quadratic function?

31. What is the range of the square root function?

Chapter 1 Test

1. Which of the following graphs is a function?

(A)

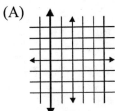

(B)

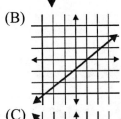

(C)

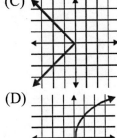

(D)

2. The function rule is $3x(x+5)$.

x	$f(x)$
-2	?

(A) -18
(B) 18
(C) -42
(D) 42

3. What is the range of the following relation?

$\{(1,2)(4,9)(7,8)(10,13)\}$

(A) $\{1,4,7,10\}$
(B) $\{2,9,8,13\}$
(C) $\{3,13,15,23\}$
(D) $\{1,3,1,3\}$

4. Which function is not linear?

(A) $f(x) = -3x - 7$
(B) $f(x) = 8$
(C) $f(x) = \frac{2}{3}x - 5$
(D) $f(x) = 2x^2$

5. Which function does not have a zero?

(A) $f(x) = x - 3$
(B) $f(x) = -7$
(C) $f(x) = 2x + 3$
(D) $f(x) = -3x + 2$

6. Which of the following relations is a function?

(A) $\{(0,-1)(1,-1)(0,-2)\}$
(B) $\{(-1,1)(-1,-1)(0,0)\}$
(C) $\{(2,1)(1,0)(0,-1)\}$
(D) $\{(-1,1)(-1,0)(-1,-1)\}$

7. Find the range of the following function for the domain $\{-2,-1,0,3\}$.

$y = \dfrac{2+x}{4}$

(A) $\left\{0, \dfrac{3}{4}, 1, \dfrac{5}{4}\right\}$

(B) $\left\{0, \dfrac{1}{4}, \dfrac{1}{2}, \dfrac{5}{4}\right\}$

(C) $\left\{1, -\dfrac{1}{4}, \dfrac{1}{2}, \dfrac{5}{4}\right\}$

(D) $\left\{\dfrac{1}{4}, \dfrac{3}{4}, \dfrac{1}{2}, \dfrac{5}{4}\right\}$

8. For $f(x) = 3x^2 - 5x$, find $f(-3)$.

(A) 12
(B) -6
(C) 42
(D) 3

9. What is the next number in this sequence?

0.03, 0.12, 0.48, 1.92, _____

(A) 1.95
(B) 3.36
(C) 5.08
(D) 7.68

10. What is the next number in this sequence?

2, 16, 128, 1024, _____

(A) 1,920
(B) 8,192
(C) 11,586
(D) 16,384

11. Which function has only one vertex?

(A) the linear function
(B) the quadratic function
(C) the cubic function
(D) the square root function

12. A radical is always in what kind of function?

(A) the linear function
(B) the absolute value function
(C) the square root function

(D) $\dfrac{1}{x}$

13. $f(x) = ax^3 + bx^2 + cx + d$ is the standard form of which function?

(A) the cubic function
(B) the square root function
(C) the quadratic function
(D) the linear function

14. Which function has a V shape?

(A) the linear function
(B) the absolute value function
(C) the cubic function
(D) the square root function

15. Which function is undefined at $x = 0$?

(A) the cubic function
(B) the square root function
(C) the absolute value function

(D) $\dfrac{1}{x}$

16. Which equation is a function of x?

(A) $f(b) = 3b + 2$
(B) $f(x) = 3x + 2$
(C) $f(a) = 3a + 2$
(D) $f(z) = 3z + 2$

17. Which is the equation $y = 7x^2 + 2x + 3$ as a function of m?

(A) $m = 7x^2 + 2x + 3$
(B) $y = 7m^2 + 2m + 3$
(C) $m(x) = 7m^2 + 2m + 3$
(D) $f(m) = 7m^2 + 2m + 3$

Chapter 2
Graphing Functions

This chapter covers the following Georgia Performance Standards:

MA1A	Algebra	MA1A1b
		MA1A1c
		MA1A1d
		MA1A1e
		MA1A1g

2.1 Graphing Linear Equations

In addition to graphing ordered pairs, the Cartesian plane can be used to graph the solution set for an equation. Any equation with two variables that are both to the first power is called a **linear equation.** The graph of a linear equation will always be a straight line.

Example 1: Graph the solution set for $x + y = 7$.

Step 1: Make a list of some pairs of numbers that will work in the equation.

$$\frac{x + y = 7}{\begin{array}{ll} 4 + 3 = 7 & (4, 3) \\ -1 + 8 = 7 & (-1, 8) \\ 5 + 2 = 7 & (5, 2) \\ 0 + 7 = 7 & (0, 7) \end{array}}\Bigg\} \text{ ordered pair solutions}$$

Step 2: Plot these points on a Cartesian plane.

Step 3: By passing a line through these points, we graph the solution set for $x + y = 7$. This means that every point on the line is a solution to the equation $x + y = 7$. For example, $(1, 6)$ is a solution, so the line passes through the point $(1, 6)$.

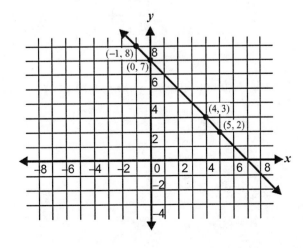

Make a table of solutions for each linear equation below. Then plot the ordered pair solutions on graph paper. Draw a line through the points. (If one of the points does not line up, you have made a mistake.)

1. $x + y = 6$

2. $y = x + 1$

3. $y = x - 2$

4. $x + 2 = y$

5. $x - 5 = y$

6. $x - y = 0$

Example 2:   Graph the equation $y = 2x - 5$.

Step 1:   This equation has 2 variables, both to the first power, so we know the graph will be a straight line. Substitute some numbers for x or y to find pairs of numbers that satisfy the equation. For the above equation, it will be easier to substitute values of x in order to find the corresponding value for y. Record the values for x and y in a table.

x	y
0	-5
1	-3
2	-1
3	1

If x is 0, y would be -5
If x is 1, y would be -3
If x is 2, y would be -1
If x is 3, y would be 1

Step 2:   Graph the ordered pairs, and draw a line through the points.

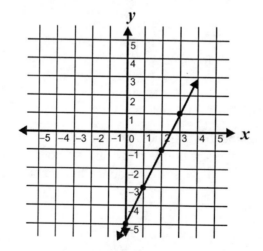

Find pairs of numbers that satisfy the equations below, and graph the line on graph paper.

1. $y = -2x + 2$

2. $2x - 2 = y$

3. $-x + 3 = y$

4. $y = x + 1$

5. $4x - 2 = y$

6. $y = 3x - 3$

7. $x = 4y - 3$

8. $2x = 3y + 1$

9. $x + 2y = 4$

2.2   Slope-Intercept Form of a Line

An equation that contains two variables, each to the first degree, is a **linear equation**. The graph for a linear equation is a straight line. To put a linear equation in slope-intercept form, solve the equation for y. This form of the equation shows the slope and the y-intercept. Slope-intercept form follows the pattern of $y = mx + b$. The "m" represents slope, and the "b" represents the y-intercept. The y-intercept is the point at which the line crosses the y-axis. The constant is called the **slope** of the line.

Example 3:   Put the equation $2x + 3y = 15$ in slope-intercept form. What is the slope of the line? What is the y-intercept? Graph the line.

Step 1:   Solve for y:

$$
\begin{array}{rcr}
2x + 3y &=& 15 \\
-2x & & -2x \\
\hline
\dfrac{3y}{3} &=& -\dfrac{2x}{3} + \dfrac{15}{3}
\end{array}
$$

slope-intercept form: $y = -\frac{2}{3}x + 5$

The slope is $-\frac{2}{3}$ and the y-intercept is 5.

Step 2:   Knowing the slope and the y-intercept, we can graph the line.

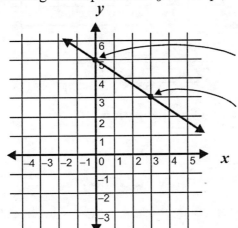

The y-intercept is 5, so the line passes through the point $(0, 5)$ on the y-axis.

The slope is $-\frac{2}{3}$, so go down 2 and over 3 to get a second point.

Put each of the following equations in slope-intercept form by solving for y. On your graph paper, graph the line using the slope and y-intercept.

1. $4x - y = 5$
2. $2x + 4y = 16$
3. $3x - 2y = 10$
4. $x + 3y = -12$
5. $6x + 2y = 0$

6. $8x - 5y = 10$
7. $-2x + y = 4$
8. $-4x + 3y = 12$
9. $-6x + 2y = 12$
10. $x - 5y = 5$

11. $3x - 2y = -6$
12. $3x + 4y = 2$
13. $-x = 2 + 4y$
14. $2x = 4y - 2$
15. $6x - 3y = 9$

16. $4x + 2y = 8$
17. $6x - y = 4$
18. $-2x - 4y = 8$
19. $5x + 4y = 16$
20. $6 = 2y - 3x$

2.3   Graphing Basic Functions

A graph is an image that shows the relationship between two or more variables. In this section, we will learn how to graph six basic functions.

Example 4: $f(x) = x$

The notation $f(x)$ is the same as y, it just means f as a function of x

The easiest way to begin graphing this is to draw a table of values. To create an accurate graph, you should choose at least 5 values for your table. These values are now your (x, y) ordered pair and they are ready to be plotted.

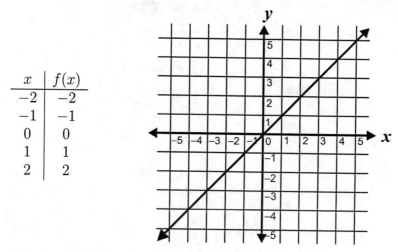

x	$f(x)$
-2	-2
-1	-1
0	0
1	1
2	2

This is called the graph of a **linear function**.

Example 5: $f(x) = x^2$

Exponential functions can be graphed the same way as linear functions.

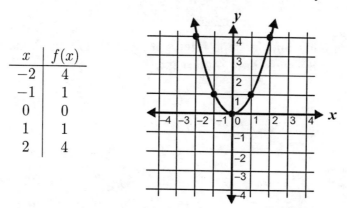

x	$f(x)$
-2	4
-1	1
0	0
1	1
2	4

This is called the graph of a **quadratic function**.

Example 6: $f(x) = x^3$

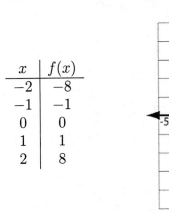

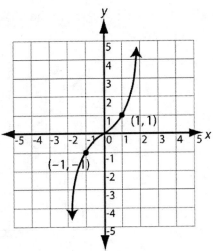

x	$f(x)$
-2	-8
-1	-1
0	0
1	1
2	8

This is called the graph of a **cube root function**.

Example 7: $f(x) = \sqrt{x}$

An equation like this will be easier to graph if you choose values of x that are perfect squares and because you can't take the square root of a negative, there is no need to select negative values.

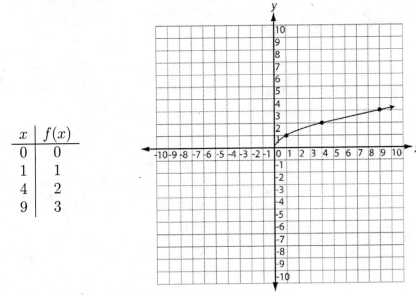

x	$f(x)$
0	0
1	1
4	2
9	3

This is called the graph of a **square root function**.

Example 8: $f(x) = |x|$

Don't forget, when you take the absolute value of a negative number, it becomes positive!

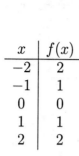
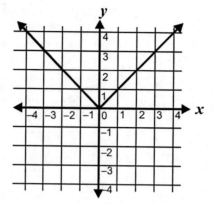

x	$f(x)$
-2	2
-1	1
0	0
1	1
2	2

This is called the graph of an **absolute value function**.

Example 9: $f(x) = \dfrac{1}{x}$

Don't forget, anything over zero is undefined and therefore does not exist!

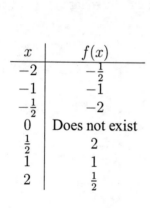
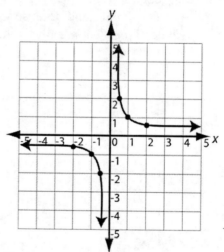

x	$f(x)$
-2	$-\frac{1}{2}$
-1	-1
$-\frac{1}{2}$	-2
0	Does not exist
$\frac{1}{2}$	2
1	1
2	$\frac{1}{2}$

Graph the following.

1. $f(x) = 2x$

2. $f(x) = 3x^3$

3. $f(x) = -\left|\frac{1}{2}x\right|$

4. $f(x) = |4x|$

5. $f(x) = \sqrt{2x}$

6. $f(x) = \dfrac{6}{x}$

7. $f(x) = \frac{1}{3}x^2$

8. $f(x) = -4x^2$

9. $f(x) = -\dfrac{2}{x}$

10. $f(x) = -\sqrt{x}$

2.4   Understanding Slope

The slope of a line refers to how steep a line is. Slope is also defined as the rate of change. When we graph a line using ordered pairs, we can easily determine the slope. Slope is often represented by the letter m.

The formula for slope of a line is: $m = \dfrac{y_2 - y_1}{x_2 - x_1}$ or $\dfrac{\text{rise}}{\text{run}}$

Example 10:   What is the slope of the following line that passes through the ordered pairs $(-4, -3)$ and $(1, 3)$?

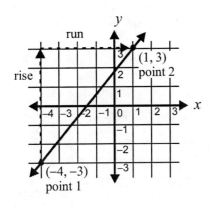

y_2 is 3, the y-coordinate of point 2.

y_1 is -3, the y-coordinate of point 1.

x_2 is 1, the x-coordinate of point 2.

x_1 is -4, the x-coordinate of point 1.

Use the formula for slope given above:

$$m = \frac{3 - (-3)}{1 - (-4)} = \frac{6}{5}$$

The slope is $\frac{6}{5}$. This shows us that we can go up 6 (rise) and over 5 to the right (run) to find another point on the line.

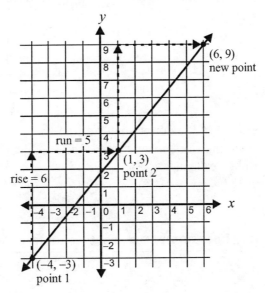

Example 11:   Find the slope of a line through the points $(-2, 3)$ and $(1, -2)$. It doesn't matter which pair we choose for point 1 and point 2. The answer is the same.

Let point 1 be $(-2, 3)$
Let point 2 be $(1, -2)$

$$\text{slope} = \frac{(y_2 - y_1)}{(x_2 - x_1)} = \frac{-2 - 3}{1 - (-2)} = \frac{-5}{3}$$

When the slope is negative, the line will slant up to the left. For this example, the line will go **down** 5 units and then over 3 units to the **right**.

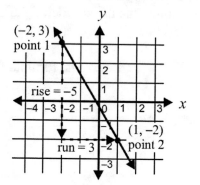

Example 12:   What is the slope of a line that passes through $(1, 1)$ and $(3, 1)$?

$$\text{slope} = \frac{1 - 1}{3 - 1} = \frac{0}{2} = 0$$

When $y_2 - y_1 = 0$, the slope will equal 0, and the line will be horizontal.

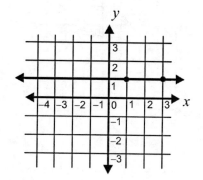

Example 13:   What is the slope of a line that passes through $(2, 1)$ and $(2, -3)$?

$$\text{slope} = \frac{-3 - 1}{2 - 2} = \frac{-4}{0} = \text{undefined}$$

When $x_2 - x_1 = 0$, the denominator of the slope is 0, then the slope is undefined, and the line will be vertical.

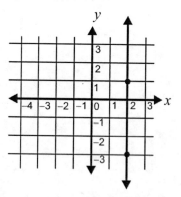

The following lines summarize what we know about slope.

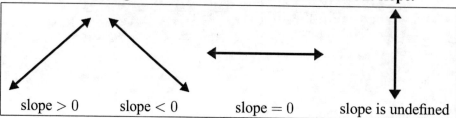

slope > 0   slope < 0   slope = 0   slope is undefined

Find the slope of the line that goes through the following pairs of points. Then, using graph paper, graph the line through the two points, and label the rise and run. (See Examples 10–13).

1. $(2, 3)$ $(4, 5)$

2. $(1, 3)$ $(2, 5)$

3. $(-1, 2)$ $(4, 1)$

4. $(1, -2)$ $(4, -2)$

5. $(3, 0)$ $(3, 4)$

6. $(3, 2)$ $(-1, 8)$

7. $(4, 3)$ $(2, 4)$

8. $(2, 2)$ $(1, 5)$

9. $(3, 4)$ $(1, 2)$

10. $(3, 2)$ $(3, 6)$

11. $(6, -2)$ $(3, -2)$

12. $(1, 2)$ $(3, 4)$

13. $(-2, 1)$ $(-4, 3)$

14. $(5, 2)$ $(4, -1)$

15. $(1, -3)$ $(-2, 4)$

16. $(2, -1)$ $(3, 5)$

2.5 Comparing Rates of Change

The rate of change is the value at which the quantity, y, is changing in respect to quantity, x.

$$\text{Rate of Change} = \text{change in } y \div \text{change in } x \left(\frac{y_2 - y_1}{x_2 - x_1} \right)$$

In a linear function, the rate of change, also known as the slope, remains constant. This is great for describing an object moving at a constant speed, but that is not always the case. That is why, in quadratic functions, the rate of change varies. Quadratic equations have a variable rate of change.

Example 14:   Linear Functions

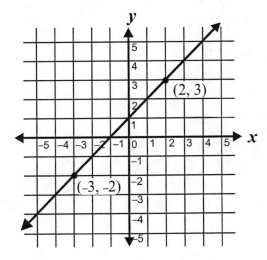

Find the slope: $\dfrac{y_2 - y_1}{x_2 - x_1} = \dfrac{3 - (-2)}{2 - (-3)} = \dfrac{5}{5} = 1$

This value shows that every time x increases by one, y does the same. The slope of a linear equation is always constant.

Example 15:   Quadratic Function, $y = x^2$

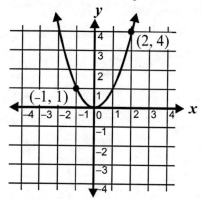

Find the slope: $\dfrac{4-1}{2-(-1)} = \dfrac{3}{3} = 1$

This value can change depending on the points you choose, so the slope of a quadratic equation is variable.

Example 16:   Absolute Value Function $y = |x|$

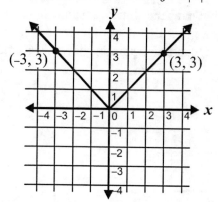

Find the slope: $\dfrac{3-3}{3-(-3)} = \dfrac{0}{6} = 0$

This, like the quadratic function, is constantly changing depending on the points you choose. It has a variable rate of change.

Graph and tell whether the slope is constant (c) or variable (v).

1. $f(x) = 3x + 2$
2. $f(x) = -x^2 + 4$
3. $f(x) = |x| + 7$
4. $f(x) = \frac{1}{2}x^2 - 1$
5. $f(x) = \sqrt{x}$

2.6   Parent Functions

The simplest function is a straight line, one in which the equation that represents it never has variables raised to a power other than 1. The simplest linear function, $y = x$, is called the **linear parent function**. Another way of saying this is that if the graph of any function is a straight line, then its parent function is $y = x$.

Example 17:   What is the parent function of the following graph?

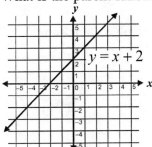

Since the graph is a straight line, the parent function is $y = x$, the linear parent function.

If an equation can be written in the form $y = ax^2 + bx + c$, where a, b, and c are constants and $a \neq 0$, then it is a **quadratic equation**. If a quadratic can be written in this form, then its **quadratic parent function** is $y = x^2$. Another way of stating this is that if the graph of any function is a parabola as shown, then its parent function is $y = x^2$.

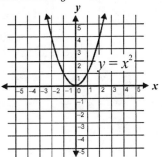

Example 18:   What is the parent function of this graph?

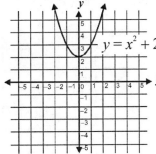

Since the graph is a parabola, its parent function is the quadratic parent function $y = x^2$. We also will notice that the equation is $y = x^2 + 2$. By the definition, this equation is quadratic, making its parent function $y = x^2$.

Name the parent function of the following equations.

1. $y = -\frac{1}{2}x + 5$ 3. $y = x^2 - 4$ 5. $y = 4x - 2$ 7. $y = x^2 + x + \frac{3}{2}$

2. $y = 10x$ 4. $y = -\frac{1}{3}x^2 + 8$ 6. $y = (x - 2)^2 - 1$ 8. $y = 17 - 4x$

2.7   Transformations of Graphs

Graph transformation occurs when a graph is altered from its original or "parent" state. Graphs can be altered in 5 ways: vertical shifts, horizontal shifts, they can be stretched, shrunk, or reflected.

$y = f(x + a) + b$

a = horizontal shift (left/right), b = vertical shift (up/down)

Shifting Upward and Downward

Example 19:   Given the graph of $f(x)$, what is the graph of function $y = f(x) + 3$?

$f(x) = x, y = x + 3$

$f(x) = x^2, y = x^2 + 3$

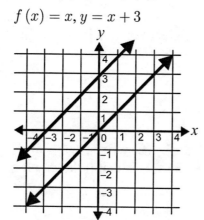 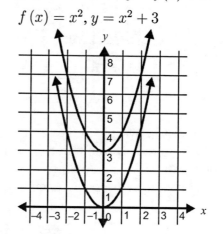

The graph of function $y = f(x) + 3$ shifts the graph of $f(x)$ up 3 units.

This transformation is called a **vertical shift**.

Shifting Left and Right

Example 20:   Given the graph of $f(x)$, what is the graph of function $y = f(x + 3)$?

$f(x) = x^2, y = (x + 3)^2$

$f(x) = |x|, y = |x + 3|$

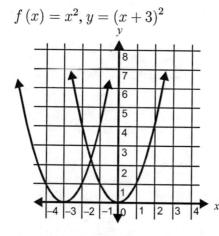 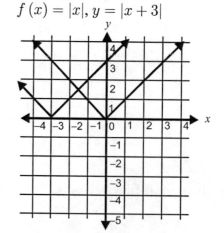

The graph of function $y = f(x + 3)$ shifts the graph of $f(x)$ left 3 units.

This transformation is called a **horizontal shift**.

Vertical Stretching and Shrinking

Example 21:   Given the graph of $f(x)$, what is the graph of function $y = 3f(x)$?

$f(x) = x^2, y = 3x^2$

$f(x) = |x|, y = 3|x|$

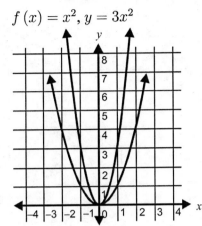 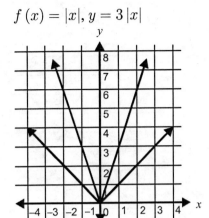

The graph of function $y = 3f(x)$ stretches the graph of $f(x)$ vertically.

This transformation is called **stretching**.

Note: If $c > 1$, then the graph of $y = cf(x)$ is the graph of $f(x)$ stretched vertically by the ratio of c. If $0 < c < 1$, then the graph of $y = cf(x)$ is the graph of $f(x)$ compressed (shrunk) vertically by the ratio of c.

Horizontal Stretching and Shrinking

Example 22:   Given the graph of $f(x)$, what is the graph of function $y = f(3x)$?

$f(x) = x^2, y = (3x)^2 = 9x^2$

$f(x) = x^2 + 1, y = 9x^2 + 1$

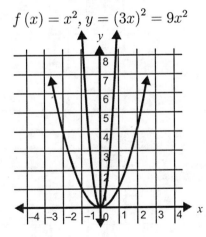 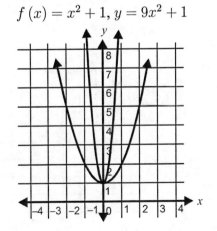

The graph of function $y = f(3x)$ shrinks the graph of $f(x)$ horizontally.

This transformation is called **shrinking**.

Note: If $c > 1$, then the graph of $y = f(cx)$ is the graph of $f(x)$ shrunk horizontally by the ratio of c. If $0 < c < 1$, then the graph of $y = f(cx)$ is the graph of $f(x)$ stretched horizontally by the ratio of c.

Reflection across the x-axis

Example 23:   Given the graph of $f(x)$, what is the graph of function $y = -f(x)$?

$f(x) = x,\ y = -x$

$f(x) = x^2,\ y = -x^2$

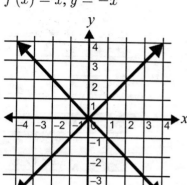
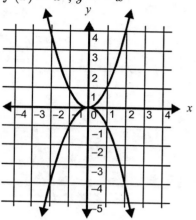

The graph of function $y = -f(x)$ reflects the graph of $f(x)$ across the x-axis. This transformation is called a **reflection**.

Reflection across the y-axis

Example 24:   Given the graph of $f(x)$, what is the graph of function $y = f(-x)$?

$f(x) = \sqrt{x},\ y = \sqrt{-x}$

$f(x) = x^3,\ y = (-x)^3 = -x^3$

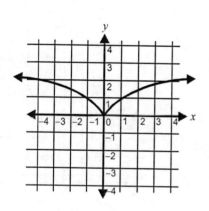
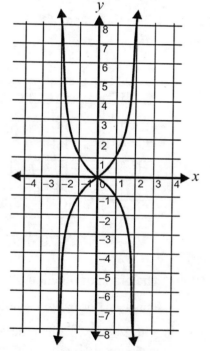

The graph of function $y = f(-x)$ reflects the graph of $f(x)$ across the y-axis. This transformation is called a **reflection**.

Complete the following.

1. How is $y = x - 7$ different from its parent graph?

2. How does $y = 3x^2$ differ from its parent graph?

3. What is the parent graph of $y = x^2 - 12$?

4. Does $y = -f(x)$ reflect across the x or y-axis?

5. Does $y = f(-x)$ reflect across the x or y-axis?

6. How would adding 3 to $y = x^2$ change the graph?

7. How does subtracting 3 from $y = x^2$ change the graph?

8. Would $y = (x + 3)^2$ be shifted vertically or horizontally from its parent function?

9. Would $y = f(3x)$ shrink the graph of $f(x)$ horizontally or vertically?

10. What is the parent function of $y = x + 12$?

11. What is the parent function of $y = x^2 + 3x - 4$?

12. Would $y = |x| - 7$ be shifted vertically or horizontally from its parent function?

2.8   Determining Domain and Range From Graphs

The domain is all of the x values that lie on the function in the graph from the lowest x value to the highest x value. The range is all of the y values that lie on the function in the graph from the lowest y to the highest y.

Example 25:   Find the domain and range of the graph.

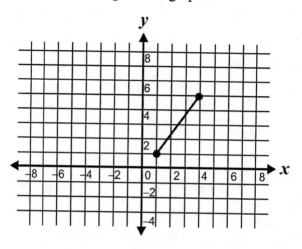

Step 1:   First find the lowest x value depicted on the graph. In this case it is 1. Then find the highest x value depicted on the graph. The highest value of x on the graph is 4. The domain must contain all of the values between the lowest x value and the highest x value. The easiest way to write this is $1 \leq$ Domain ≤ 4 or $1 \leq x \leq 4$.

Step 2:   Perform the same process for the range, but this time look at the lowest and highest y values. The answer is $1 \leq$ Range ≤ 5 or $1 \leq y \leq 5$.

Find the domain and range of each graph below. Write your answers on the line provided.

1.

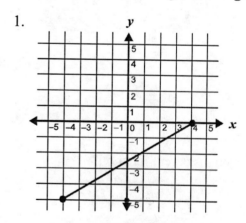

2.

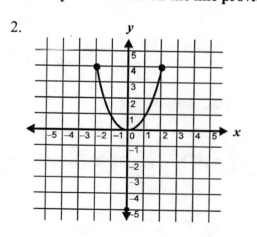

_____

_____

3.

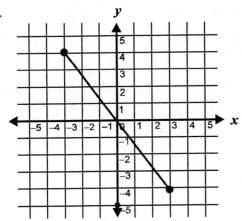

_____

4.

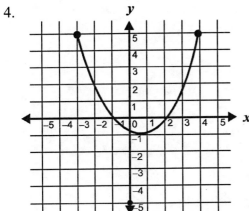

_____

5.

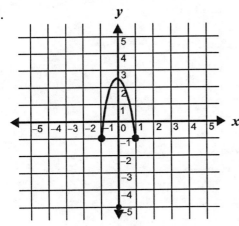

_____

6.

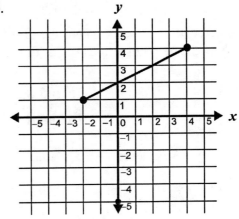

_____

7.

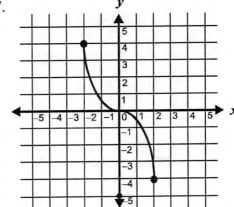

_____

8.

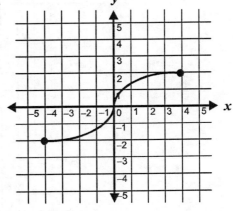

_____

2.9   Finding the Intercepts of a Line

The x-intercept is the point where the graph of a line crosses the x-axis. The y-intercept is the point where the graph of a line crosses the y-axis.

To find the x-intercept, set $y = 0$

To find the y-intercept, set $x = 0$

Example 26:   Find the x- and y-intercepts of the line $6x + 2y = 18$

Step 1:   To find the x-intercept, set $y = 0$.

$$6x + 2(0) = 18$$
$$\frac{6x}{6} = \frac{18}{6}$$
$$x = 3$$

The x-intercept is at the point $(3, 0)$.

Step 2:   To find the y-intercept, set $x = 0$.

$$6(0) + 2y = 18$$
$$\frac{2y}{2} = \frac{18}{2}$$
$$y = 9$$

The y-intercept is at the point $(0, 9)$.

Step 3:   You can now use the two intercepts to graph the line.

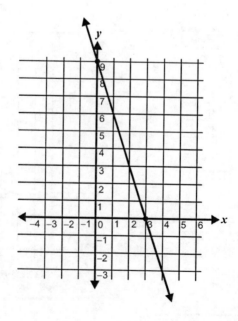

For each of the following equations, find both the x and the y intercepts of the line. For extra practice, draw each of the lines on graph paper.

1. $8x - 2y = 8$

2. $4x + 8y = 16$

3. $3x + 3y = 9$

4. $x - 2y = -5$

5. $8x + 4y = 32$

6. $3x - 4y = 12$

7. $-3x - 3y = 6$

8. $-6x + 2y = 18$

9. $4x - 2y = -4$

10. $-5x - 3y = 15$

11. $3x - 6y = -12$

12. $6x + 3y = 9$

13. $-2x - 6y = 18$

14. $2x + 3y = -6$

15. $-3x + 8y = 12$

16. $5x + 10y = 30$

17. $x - 3y = -2$

18. $-3x - 4y = 6$

19. $7x + 2y = 14$

20. $5x + 6y = 30$

21. $7x - 2y = 14$

22. $5x + 5y = 25$

23. $-14x + 4y = 28$

24. $22x - 11y = -44$

25. $-3x - 6y = -9$

26. $4x + 6y = 24$

27. $10x - 5y = 40$

28. $30x + 15y = 60$

29. $-9x + 18y = 54$

30. $8x + 7y = 56$

Chapter 2 Review

1. Graph the solution set for the linear equation: $x - 3 = y$.

2. What is the slope of the line that passes through the points $(5, 3)$ and $(6, 1)$?

3. What is the slope of the line that passes through the points $(-1, 4)$ and $(-6, -2)$?

4. What is the x-intercept for the following equation? $6x - y = 30$

5. What is the y-intercept for the following equation? $4x + 2y = 28$

6. What is the slope of the line $y = -\frac{1}{2}x + 3$?

7. What is the x-intercept of the line $y = 5x + 6$?

8. What is the y-intercept of the line $y - \frac{2}{3}x + 3 = 0$?

Graph each equation and state whether the slope is constant (c) or varying (v).

9. $f(x) = x - 6$

10. $f(x) = \frac{1}{2}x + 6$

11. $f(x) = \frac{1}{2}x^2 + 6$

12. $f(x) = \frac{1}{2}|x| + 6$

Use what you know about graph transformations to answer the following questions.

13. What is the parent graph of $y = 3x + 2$?

14. True or False: $y = f(-x)$ is reflected across the y-axis.

15. True or False: $y = 3f(x)$ shrinks a graph horizontally.

16. Write the equation of $y = x^2$ shifted down two units.

17. Write the equation of $y = x$ shifted left six units.

18. True or false: $y = f(2x)$ shrinks a graph horizontally.

Graph each function.

19. $f(x) = -|2x|$

20. $f(x) = \frac{1}{2}x^2$

21. $f(x) = |-6x|$

22. $f(x) = \frac{1}{7}x$

23. $f(x) = |4x|$

24. $f(x) = -3\sqrt{x}$

25. $f(x) = \frac{1}{4}x^3$

26. $f(x) = -\frac{1}{8}\sqrt{x}$

Chapter 2 Test

1. The following graph depicts the height of a projectile as a function of time.

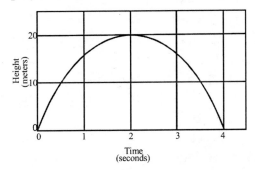

What is the domain (D) of this function?

(A) 0 meters $\leq$ D $\leq$ 20 meters
(B) 4 seconds $\leq$ D $\leq$ 20 meters
(C) 20 meters $\leq$ D $\leq$ 4 seconds
(D) 0 seconds $\leq$ D $\leq$ 4 seconds

2. What is the x-intercept of the following linear equation $3x + 4y = 12$?

(A) $(0, 3)$
(B) $(3, 0)$
(C) $(0, 4)$
(D) $(4, 0)$

3. Which of the following equations is represented by the graph?

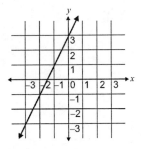

(A) $y = -3x + 3$

(B) $y = -\frac{1}{3}x + 3$

(C) $y = 3x - 3$

(D) $y = 2x + 3$

4. Which of the following is the graph of the equation $y = x - 3$?

(A)
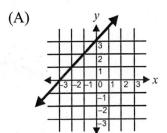

(B)
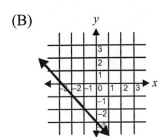

(C)
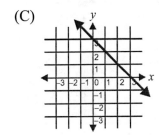

(D)
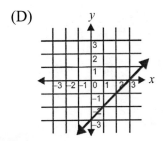

5. What function would have a variable rate of change?

(A) $f(x) = 2x^2 + 1$
(B) $f(x) = 3x - 1$
(C) $f(x) = 6x + 4$
(D) $f(x) = 9x - \frac{1}{2}$

6. What is the x-intercept and y-intercept for the equation $x + 2y = 6$?

(A) x-intercept = $(0, 6)$
y-intercept = $(3, 0)$

(B) x-intercept = $(4, 1)$
y-intercept = $(2, 2)$

(C) x-intercept = $(0, 6)$
y-intercept = $(0, 3)$

(D) x-intercept = $(6, 0)$
y-intercept = $(0, 3)$

7. Look at the graphs below. Which of the following statements is false?

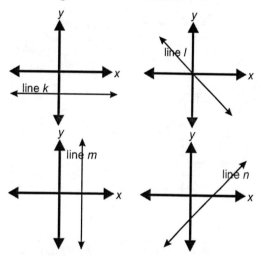

(A) The slope of line k is undefined.
(B) The slope of line l is negative.
(C) The slope of line m is undefined.
(D) The slope of line n is positive.

8. What is the rate of change of $f(x) = 3x - 6$?

(A) -6

(B) 3

(C) $\frac{1}{3}$

(D) Not enough information

9. What function would have a constant rate of change?

(A) $f(x) = 2x^2 + 1$
(B) $f(x) = 7x^3 + 1$
(C) $f(x) = 403x + 6$
(D) $f(x) = 17x^4 + 4$

10. What type of rate change does $f(x) = 2x^2 + 4$ have?

(A) variable
(B) constant
(C) increasing
(D) decreasing

11. In which direction is $y = x + 6$ shifted from its parent function?

(A) right
(B) up
(C) down
(D) none of the above

12. If you have the equation $y = f(x)$, and it is changed to $y = f(-x)$, then the graph is reflected. The graph is reflected over the

(A) y-axis.
(B) x-axis.
(C) origin.
(D) line $y = x$.

13. In the equation $y = 6f(x)$, 6 would _____ the graph.

(A) vertically stretch
(B) vertically shrink
(C) vertically shift
(D) reflect

14. In the equation $y = f(6x)$, 6 would _____ the graph.

(A) horizontally stretch
(B) horizontally shrink
(C) horizontally shift
(D) reflect

15. In the equation $y = f(x+3)$ would _____ the graph _____ units _____.

   (A) stretch, 3, right
   (B) stretch, 3, left
   (C) shift, 3, right
   (D) shift, 3, left

16. $f(x) = \dfrac{1}{x}$ does not exist at what point?

   (A) $x = 1$
   (B) $x = -1$
   (C) $x = 0$
   (D) x exists everywhere

17. Which is the graph of $f(x) = \dfrac{1}{x}$?

(A)

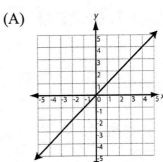

(B)

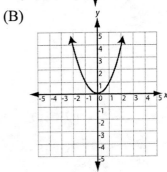

(C)

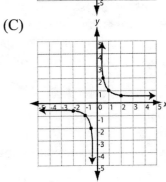

(D)

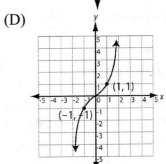

Chapter 3
Logic and Geometric Proofs

This chapter covers the following Georgia Performance Standards:

MA1G	Geometry	MA1G2a
		MA1G2b

3.1   Mathematical Reasoning/Logic

The Georgia mathematics curriculum calls for skill development in mathematical **reasoning** or **logic**. The ability to use logic is an important skill for solving math problems, but it can also be helpful in real-life situations. For example, if you need to get to Park Street, and the Park Street bus always comes to the bus stop at 3 PM, then you know that you need to get to the bus stop by at least 3 PM. This is a real-life example of using logic, which many people would call "common sense."

There are many different types of statements which are commonly used to describe mathematical principles. However, using the rules of logic, the truth of any mathematical statement must be evaluated. Below is a list of tools used in logic to evaluate mathematical statements.

Logic is the discipline that studies valid reasoning. There are many forms of valid arguments, but we will review just a few here.

A **proposition** is usually a declarative sentence which may be true or false.

An **argument** is a set of two or more related propositions, called **premises**, that provide support for another proposition, called the **conclusion**.

Deductive reasoning is an argument which begins with general premises and proceeds to a more specific conclusion. Most elementary mathematical problems use deductive reasoning.

Inductive reasoning is an argument in which the truth of its premises make it likely or probable that its conclusion is true.

3.2 Arguments

Most of logic deals with the evaluation of the validity of arguments. An argument is a group of statements that includes a conclusion and at least one premise. A premise is a statement that you know is true or at least you assume to be true. Then, you draw a conclusion based on what you know or believe is true in the premise(s). Consider the following example:

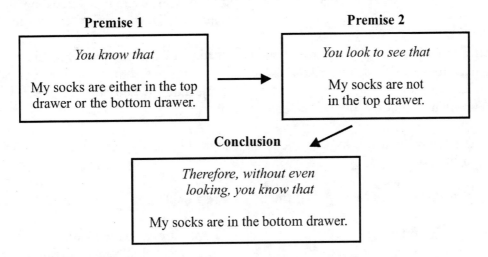

This argument is an example of deductive reasoning, where the conclusion is "deduced" from the premises and nothing else. In other words, if Premise 1 and Premise 2 are true, you don't even need to look in the bottom drawer to know that the conclusion is true.

3.3 Deductive and Inductive Arguments

In general, there are two types of logical arguments: **deductive** and **inductive**. Deductive arguments tend to move from general statements or theories to more specific conclusions. Inductive arguments tend to move from specific observations to general theories.

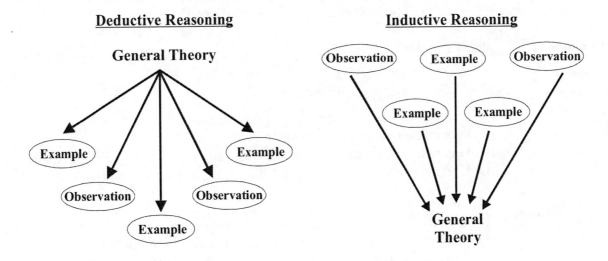

Compare the two examples below:

Deductive Argument		**Inductive Argument**	
Premise 1	All men are mortal.	**Premise 1**	The sun rose this morning.
Premise 2	Socrates is a man.	**Premise 2**	The sun rose yesterday morning.
Conclusion	Socrates is mortal.	**Premise 3**	The sun rose two days ago.
		Premise 4	The sun rose three days ago.
		Conclusion	The sun will rise tomorrow.

An inductive argument cannot be proved beyond a shadow of a doubt. For example, it's a pretty good bet that the sun will come up tomorrow, but the sun not coming up presents no logical contradiction. An inductive argument is a type of conjecture. A **conjecture** appears to be true, but has not been proven logically. It is an educated guess.

On the other hand, a deductive argument can have logical certainty, but it must be properly constructed. Consider the examples below.

True Conclusion from an Invalid Argument	**False Conclusion from a Valid Argument**
All men are mortal. Socrates is mortal. Therefore Socrates is a man.	All astronauts are men. Julia Roberts is an astronaut. Therefore, Julia Roberts is a man.
Even though the above conclusion is true, the argument is based on invalid logic. Both men and women are mortal. Therefore, Socrates could be a woman.	In this case, the conclusion is false because the premises are false. However, the logic of the argument is valid because *if* the premises were true, then the conclusion would be true.

A **counterexample** is an example in which the statement is true but the conclusion is false when we have assumed it to be true. If we said, "All cocker spaniels have blonde hair," then a counterexample would be a red-haired cocker spaniel. If we made the statement, "If a number is greater than 10, it is less than 20," we can easily think of a counterexample, like 35.

Example 1:   Which argument is valid?

If you speed on Hill Street, you will get a ticket.
If you get a ticket, you will pay a fine.

(A) I paid a fine, so I was speeding on Hill Street.
(B) I got a ticket, so I was speeding on Hill Street.
(C) I exceeded the speed limit on Hill Street, so I paid a fine.
(D) I did not speed in Hill Street, so I did not pay a fine.

Solution: C is valid.
A is incorrect. I could have paid a fine for another violation.
B is incorrect. I could have gotten a ticket for some other violation.
D is incorrect. I could have paid a fine for speeding somewhere else.

Example 2:   Assume the given proposition is true. Then determine if each statement is true or false.

Given: If a dog is thirsty, he will drink.

(A)	If a dog drinks, then he is thirsty.	T or F
(B)	If a dog is not thirsty, he will not drink.	T or F
(C)	If a dog will not drink, he is not thirsty.	T or F

Solution: A is false. He is not necessarily thirsty; he could just drink because other dogs are drinking or drink to show others his control of the water. This statement is the **converse** of the original. The converse of the statement "If A, then B" is "If B, then A."

B is false. The reasoning from A applies. This statement is the **inverse** of the original. The inverse of the statement "If A, then B" is "If not A, then not B."

C is true. It is the **contrapositive**, or the complete opposite of the original. The contrapositive says "If not B, then not A."

For numbers 1–5, what conclusion can be drawn from each proposition?

1. All squirrels are rodents. All rodents are mammals. Therefore,

2. All fractions are rational numbers. All rational numbers are real numbers. Therefore,

3. All squares are rectangles. All rectangles are parallelograms. All parallelograms are quadrilaterals. Therefore,

4. All Chevrolets are made by General Motors. All Luminas are Chevrolets. Therefore,

5. If a number is even and divisible by three, then it is divisible by six. Eighteen is divisible by six. Therefore,

For numbers 6–9, assume the given proposition is true. Then, determine if the statements following it are true or false.

All squares are rectangles.

6.	All rectangles are squares.	T or F
7.	All non-squares are non-rectangles.	T or F
8.	No squares are non-rectangles.	T or F
9.	All non-rectangles are non-squares.	T or F

3.4   Logic

A **conditional statement** is a type of logical statement that has two parts, a **hypothesis** and a **conclusion**. The statement is written in "if-then" form, where the "if" part contains the hypothesis and the "then" part contains the conclusion. Shorthand: $p \rightarrow q$. For example, let's start with the statement "Two lines intersect at exactly one point." We can rewrite this as a conditional statement in "if-then" form as follows:

$$\underbrace{\text{If two lines intersect}}_{\text{hypothesis}}, \text{ then } \underbrace{\text{their intersection is at exactly one point.}}_{\text{conclusion}}$$

Conditional statements may be true or false. To show that a statement is false, you need only to provide a single **counterexample** which shows that the statement is not always true. To show that a statement is true, on the other hand, you must show that the conclusion is true for all occasions in which the hypothesis occurs. This is often much more difficult.

Example 3:   Provide a counterexample to show that the following conditional statement is false:
If $x^2 = 4$, then $x = 2$.

To find a counterexample, think: Could x equal something else? Yes, x could be -2. $(-2)^2 = 4$
Therefore, we have provided a counterexample to show that the conditional statement is false.

The **converse** of a conditional statement is an "if-then" statement written by switching the hypothesis and the conclusion. For example, for the conditional statement "If a figure is a quadrilateral, then it is a rectangle," the converse is "If a figure is a rectangle, then it is a quadrilateral." Shorthand: $q \rightarrow p$

The **inverse** of a conditional statement is written by negating the hypothesis and conclusion of the original "if-then" conditional statement. Negating means to change the meaning so it is the negative, or opposite, of its original meaning. The inverse of the conditional statement "If a figure is a quadrilateral, then it is a rectangle" is "If a figure is **not** a quadrilateral, then it is **not** a rectangle." Shorthand: $\sim p \rightarrow \sim q$

The **contrapositive** of a conditional statement is written by negating the converse. That is, switch the hypothesis and conclusion of the original statement, and make them both negative. The contrapositive statement has the same meaning as the original statement. The contrapositive of the conditional statement "If a figure is a quadrilateral, then it is a rectangle" is "If a figure is not a rectangle, then it is not a quadrilateral." Shorthand: $\sim q \rightarrow \sim p$

Example 4: Given the conditional statement "If $m\angle F = 60°$, then $\angle F$ is acute." Write the converse, inverse and contrapositive.

Step 1: The converse (q → p) is constructed by switching the hypothesis and the conclusion: If $\angle F$ is acute, then $m\angle F = 60°$.

Step 2: The inverse (~p → ~q) is constructed by negating the original statement: If $m\angle F \neq 60°$, then $\angle F$ is not acute.

Step 3: The contrapositive (~q → ~p) is the negation of the converse: If $\angle F$ is not acute, then $m\angle F \neq 60°$.

Answer the following problems about geometry logic.

1. Rewrite the following as a conditional statement in "if-then" form: A number divisible by 8 is also divisible by 4.

2. Write the converse of the following conditional statement: If two circles have equal radii, then the circles are congruent.

3. Consider the conditional statement: If $x^4 = 81$, then $x = 3$. Is the statement true? Provide a counterexample if it is false.

4. Consider the statement: A line contains at least two points. Write as a conditional statement in "if-then" form, then write the converse, inverse, and contrapositive of the conditional statement.

5. "If a parallelogram has four congruent sides, then it is a rhombus." Write the converse, inverse, and contrapositive for the conditional statement. Which are true? Which are false?

6. "If a triangle has one right angle, then the acute angles are complementary." Write the converse, inverse, and contrapositive for the conditional statement. Indicate whether each is true or false. Can all the statements be either true or false? Explain.

7. "If a rectangle has four congruent sides, then it is a square." Write the contrapositive for the conditional statement and indicate whether it is true or false. Give a counterexample if it is false.

8. Show why a conditional statement and its inverse are always logically equivalent. Similarly, show why a statement's converse and inverse are logically equivalent.

Find the new if-then statement using the two statements below.

p = It is raining.   q = We will not go outside.

9. Find the inverse. (~p → ~q)

10. Find the converse. (q → p)

11. Find the contrapositive. (~q → ~p)

Chapter 3 Review

For numbers 1–4, assume the given proposition is true. Then determine if the statements following it are true or false.

All whales are mammals.

1. All non-whales are non-mammals.

2. If a mammal lives in the sea, it is a whale.

3. All mammals are whales.

4. All non-mammals are non-whales.

For numbers 5–8, determine whether the situation is showing deductive or inductive logic.

5. A group of students were given three descriptions about a person's job. They were then told to decide what type of job title the person has.

6. When traveling in a car on a family vacation, I noticed that I could see the ocean to my left and palm trees to my right. I concluded that my family and I were going to the beach.

7. Sammy asked her friend, Amy, to give her a good reason to get a summer job. Amy gave Sammy four good reasons to get a job.

8. The neighbor's cars are in the driveway and all of the lights in the house are off, so they must be sleeping.

Solve the following problems.

9. "If a triangle is isosceles, then its base angles are congruent." Write the contrapositive for the conditional statement. Is the statement true or false? Is the contrapositive true or false? If false, give a counterexample.

10. "If the radius of a circle is doubled, then the area of the circle is increased by a factor of four." Write the converse, inverse, and contrapositive for the conditional statement. Indicate which ones are true or false.

11. "If today is Tuesday, then it is raining." Write the converse, inverse, and contrapositive for the conditional statement. Could the statements be true? Give a counterexample to prove each statement false.

For the next three questions, find the new if-then statement about the two statements below. Then determine if the new statement is true or false. If it is false, give an explanation.

p = A figure has two sets of parallel lines. q = The figure is a parallelogram.

12. Find the inverse. ($\sim p \to \sim q$)

13. Find the converse. ($q \to p$)

14. Find the contrapositive. ($\sim q \to \sim p$)

Chapter 3 Test

For 1–3, chose which argument is valid.

1. If I oversleep, I miss breakfast. If I miss breakfast, I cannot concentrate in class. If I do not concentrate in class, I make bad grades.

   (A) I made bad grades today, so I missed breakfast.
   (B) I made good grades today, so I got up on time.
   (C) I could not concentrate in class today, so I overslept.
   (D) I had no breakfast today, so I overslept.

2. If I do not maintain my car regularly, it will develop problems. If my car develops problems, it will not be safe to drive. If my car is not safe to drive, I cannot take a trip in it.

   (A) If my car develops problems, I did not maintain it regularly.
   (B) I took a trip in my car, so I maintained it regularly.
   (C) If I maintain my car regularly, it will not develop problems.
   (D) If my car is safe to drive, it will not develop problems.

3. If two triangles have all corresponding sides and all corresponding angles congruent, then they are congruent triangles. If two triangles are congruent, then they are similar triangles.

   (A) Similar triangles have all sides and all angles congruent.
   (B) If two triangles are similar, then they are congruent.
   (C) If two triangles are not congruent, then they are not similar.
   (D) If two triangles have all corresponding sides and angles congruent, then they are similar triangles.

4. Cynthia is asked to list five duties of the President. What type of logic is Cynthia using?

   (A) mathematical reasoning
   (B) inductive reasoning
   (C) intuitive reasoning
   (D) deductive reasoning

Answer the next three questions about the two statements below.

p = A figure has four sides. q = It is a square.

5. What is the inverse, and is it true?

   (A) $\sim p \to \sim q$, yes
   (B) $\sim p \to \sim q$, no
   (C) $p \to q$, yes
   (D) $q \to p$, no

6. What is the converse, and is it true?

   (A) $q \to p$, yes
   (B) $\sim q \to \sim p$, yes
   (C) $\sim p \to \sim q$, no
   (D) $q \to p$, no

7. What is the contrapositive, and is it true?

   (A) $\sim q \to \sim p$, yes
   (B) $p \to q$, yes
   (C) $q \to p$, no
   (D) $\sim q \to \sim p$, no

Chapter 4
Coordinate Geometry

This chapter covers the following Georgia Performance Standards:

MA1G	Geometry	MA1G1a
		MA1G1b
		MA1G1c
		MA1G1d
		MA1G1e

4.1   Finding the Distance Between Two Points

Notice that a subscript added to the x and y identifies each ordered pair uniquely in the plane. For example, point 1 is identified as (x_1, y_1), point 2 as (x_2, y_2), and so on. This unique subscript identification allows us to calculate slope, distance, and midpoints of line segments in the plane using standard formulas like the distance formula. To find the distance between two points on a Cartesian plane, use the following formula:

$$d = \sqrt{(y_2 - y_1)^2 + (x_2 - x_1)^2}$$

Example 1:   Find the distance between $(-2, 1)$ and $(3, -4)$.

Plugging the values from the ordered pairs into the formula, we find:

$$d = \sqrt{(-4 - 1)^2 + [3 - (-2)]^2}$$
$$d = \sqrt{(-5)^2 + (5)^2}$$
$$d = \sqrt{25 + 25} = \sqrt{50}$$

To simplify, we look for perfect squares that are a factor of 50. $50 = 25 \times 2$. Therefore,
$$d = \sqrt{25} \times \sqrt{2} = 5\sqrt{2}$$

Find the distance between the following pairs of points using the distance formula above.

1. $(6, -1) \, (5, 2)$

2. $(-4, 3) \, (2, -1)$

3. $(10, 2) \, (6, -1)$

4. $(-2, 5) \, (-4, 3)$

5. $(8, -2) \, (3, -9)$

6. $(2, -2) \, (8, 1)$

7. $(3, 1) \, (5, 5)$

8. $(-2, -1) \, (3, 4)$

9. $(5, -3) \, (-1, -5)$

10. $(6, 5) \, (3, -4)$

11. $(-1, 0) \, (-9, -8)$

12. $(-2, 0) \, (-6, 6)$

13. $(2, 4) \, (8, 10)$

14. $(-10, -5) \, (2, -7)$

15. $(-3, 6) \, (1, -1)$

4.2   Finding the Distance Between a Point and a Line

The purpose of this section is to find the distance between a point and a line. You can use a formula to find the distance between a point and a line. The formula that gives the distance between a point (m, n) and line $ax + by + c = 0$ is

$$d = \frac{|am + bn + c|}{\sqrt{a^2 + b^2}}.$$

Example 2:   Given the line $y = 2x + 4$ and a point $(4, 1)$. Find the distance between the line and the point using the formula above.

Step 1:   First, we must rearrange the equation of the line $y = 2x + 4$ into the form $ax + by + c = 0$. To do this, subtract everything from the right to put all the terms on the left side of the equal sign.
$$y - 2x - 4 = 0$$
$$-2x + y - 4 = 0$$

Step 2:   We know from the point $(4, 1)$ and the equation $-2x + y - 4 = 0$ that $m = 4$, $n = 1$, $a = -2$, $b = 1$, and $c = -4$.
Plug these values into the formula and solve for d.

$$d = \frac{|am + bn + c|}{\sqrt{a^2 + b^2}}$$

$$d = \frac{|(-2)(4) + (1)(1) + (-4)|}{\sqrt{(-2)^2 + 1^2}}$$

$$d = \frac{|-8 + 1 + (-4)|}{\sqrt{4 + 1}}$$

$$d = \frac{|-11|}{\sqrt{5}} = \frac{11}{\sqrt{5}} = 4.92$$

The distance between the point and the line is 4.92.

Using the formula above, find the shortest distance from the points and lines listed below. Round your decimal answers to the nearest hundredth.

1. $y = 3x - 2$, $(2, 2)$

2. $y = -x + 4$, $(0, 0)$

3. $y = 2x + 2$, $(3, 5)$

4. $y = x$, $(4, 11)$

5. $y = 6x + 4$, $(4, 2)$

6. $y = -3x + 7$, $(0, 0)$

7. $y = \frac{1}{4}x$, $(0, 4)$

8. $y = -\frac{1}{2}x + 4$, $(1, 1)$

4.3   Finding the Midpoint of a Line Segment

You can use the coordinates of the endpoints of a line segment to find the coordinates of the midpoint of the line segment. The formula to find the midpoint between two coordinates is:

$$\text{midpoint, } M = \left(\frac{x_1 + x_2}{2}, \frac{y_1 + y_2}{2} \right)$$

Example 3:   Find the midpoint of the line segment having endpoints at $(-3, -1)$ and $(4, 3)$.

Use the formula for the midpoint. $M = \left(\frac{4 + (-3)}{2}, \frac{3 + (-1)}{2} \right)$

When we simplify each coordinate, we find the midpoint, M, is $\left(\frac{1}{2}, 1 \right)$.

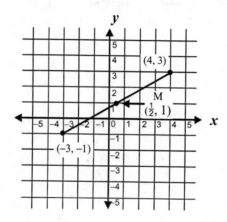

For each of the following pairs of points, find the coordinate of the midpoint, M, using the formula given above.

1. $(4, 5)\,(-6, 9)$

2. $(-3, 2)\,(-1, -2)$

3. $(3, 6)\,(9, 12)$

4. $(2, 5)\,(6, 9)$

5. $(8, 9)\,(6, 11)$

6. $(-4, 3)\,(8, 7)$

7. $(-1, -5)\,(-3, -11)$

8. $(4, 2)\,(-2, 8)$

9. $(4, 3)\,(-1, -5)$

10. $(-6, 2)\,(8, -8)$

11. $(-3, 9)\,(-9, 3)$

12. $(7, 8)\,(11, 6)$

13. $(12, 19)\,(2, 3)$

14. $(5, 4)\,(9, -2)$

15. $(-4, 6)\,(10, -2)$

4.4 Pythagorean Theorem

Pythagoras was a Greek mathematician and philosopher who lived around 600 B.C. He started a math club among Greek aristocrats called the Pythagoreans. Pythagoras formulated the **Pythagorean Theorem** which states that in a **right triangle**, the sum of the squares of the legs of the triangle are equal to the square of the hypotenuse. Most often you will see this formula written as $a^2 + b^2 = c^2$. **This relationship is only true for right triangles.**

Example 4: Find the length of side c.

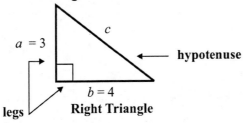

Formula:
$$a^2 + b^2 = c^2$$
$$3^2 + 4^2 = c^2$$
$$9 + 16 = c^2$$
$$25 = c^2$$
$$\sqrt{25} = \sqrt{c^2}$$
$$5 = c$$

Find the hypotenuse of the following triangles. Round the answers to two decimal places.

1.

$c =$ _____

4.

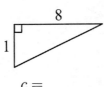

$c =$ _____

7.

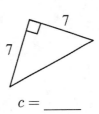

$c =$ _____

2.

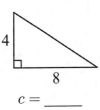

$c =$ _____

5.

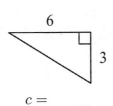

$c =$ _____

8.

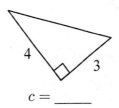

$c =$ _____

3.

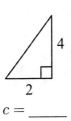

$c =$ _____

6.

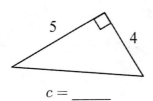

$c =$ _____

9.

$c =$ _____

4.5   Finding the Missing Leg of a Right Triangle

In some triangles, we know the measurement of the hypotenuse as well as one of the legs. To find the measurement of the other leg, use the Pythagorean theorem by filling in the known measurements, and then solve for the unknown side.

Example 5:   Find the measure of b.

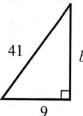

In the formula, $a^2 + b^2 = c^2$, a and b are the legs and c is always the hypotenuse.
$9^2 + b^2 = 41^2$
$81 + b^2 = 1681$
$b^2 = 1681 - 81$
$b^2 = 1600$
$\sqrt{b^2} = \sqrt{1600}$
$b = 40$

Practice finding the measure of the missing leg in each right triangle below. Simplify square roots.

1.

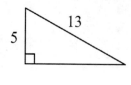

_____

4.

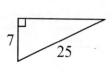

_____

7.

_____

2.

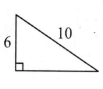

_____

5.

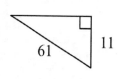

_____

8.

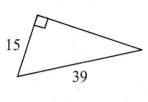

_____

3.

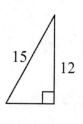

_____

6.

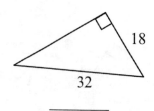

_____

9.

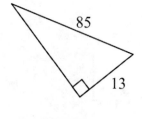

_____

4.6   Applications of the Pythagorean Theorem

The Pythagorean Theorem can be used to determine the distance between two points in some situations. Recall that the formula is written $a^2 + b^2 = c^2$.

Example 6:   Find the distance between point B and point A given that the length of each square is 1 inch long and 1 inch wide.

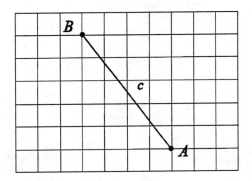

Step 1:   Draw a straight line between the two points. We will call this side c.

Step 2:   Draw two more lines, one from point B and one from point A. These lines should make a 90° angle. The two new lines will be labeled a and b. Now we can use the Pythagorean Theorem to find the distance from Point B to Point A.

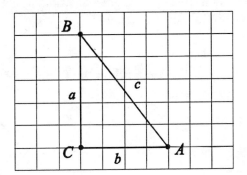

Step 3:   Find the length of a and c by counting the number of squares each line has. We find that $a = 5$ inches and $b = 4$ inches. Now, substitute the values found into the Pythagorean Theorem.

$$a^2 + b^2 = c^2$$
$$5^2 + 4^2 = c^2$$
$$25 + 16 = c^2$$
$$41 = c^2$$
$$\sqrt{41} = \sqrt{c^2}$$
$$\sqrt{41} = c$$

Use the Pythagorean Theorem to find the distances asked. Round your answers to two decimal points.

Below is a diagram of the mall. Use the grid to help answer questions 1 and 2. Each square is 25 feet × 25 feet.

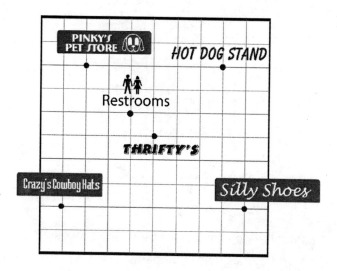

1. Marty walks from Pinky's Pet Store to the restroom to wash his hands. How far did he walk?

2. Betty needs to meet her friend at Silly Shoes, but she wants to get a hot dog first. If Betty is at Thrifty's, how far will she walk to meet her friend?

Below is a diagram of a football field. Use the grid on the football field to help find the answers to questions 3 and 4. Each square is 10 yards × 10 yards.

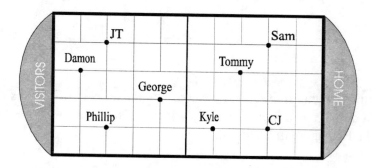

3. George must throw the football to a teammate before he is tackled. If CJ is the only person open, how far must George be able to throw the ball?

4. Damon has the football and is about to make a touchdown. If Phillip tries to stop him, how far must he run to reach Damon?

4.7 Drawing Geometric Figures on a Cartesian Coordinate Plane

You can use a Cartesian coordinate plane to draw geometric figures by plotting **vertices** and connecting them with line segments.

Example 7: What are the coordinates of each vertex of quadrilateral $ABCD$ below?

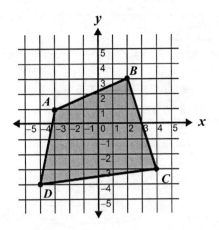

Step 1: To find the coordinates of point A, count over -3 on the x-axis and up 1 on the y-axis. point $A = (-3, 1)$.

Step 2: The coordinates of point B are located to the right two units on the x-axis and up 3 units on the y-axis. point $B = (2, 3)$.

Step 3: Point C is located 4 units to the right on the x-axis and down -3 on the y-axis. point $C = (4, -3)$.

Step 4: Point D is -4 units left on the x-axis and down -4 units on the y-axis. point $D = (-4, -4)$.

Example 8: Plot the following points. Then construct and identify the geometric figure that you plotted.

$$A = (-2, -5) \qquad B = (-2, 1) \qquad C = (3, 1) \qquad D = (3, -5)$$

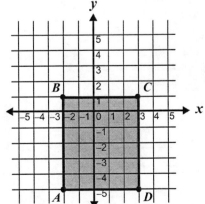

Figure $ABCD$ is a rectangle.

Find the coordinates of the geometric figures graphed below.

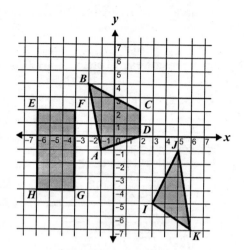

1. Quadrilateral $ABCD$
    $A =$ _____
    $B =$ _____
    $C =$ _____
    $D =$ _____

2. Rectangle $EFGH$
    $E =$ _____
    $F =$ _____
    $G =$ _____
    $H =$ _____

3. Triangle IJK
    $I =$ _____
    $J =$ _____
    $K =$ _____

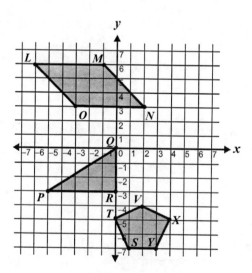

4. Parallelogram $LMNO$
    $L =$ _____
    $M =$ _____
    $N =$ _____
    $O =$ _____

5. Right Triangle PQR
    $P =$ _____
    $Q =$ _____
    $R =$ _____

6. Pentagon $STVXY$
    $S =$ _____
    $T =$ _____
    $V =$ _____
    $X =$ _____
    $Y =$ _____

Plot and label the following points. Then construct and identify the geometric figure you plotted. Question 1 is done for you.

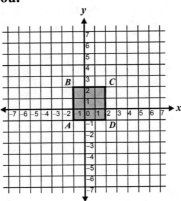

Figure   **Figure**

1. Point $A = (-1, -1)$ 3. Point $H = (-4, 0)$
    Point $B = (-1, 2)$ Point $I = (-6, 0)$
    Point $C = (2, 2)$ Point $J = (-4, 4)$
    Point $D = (2, -1)$ **square**   Point $K = (-2, 4)$ _____

2. Point $E = (3, -2)$ 4. Point $L = (-1, -3)$
    Point $F = (5, 1)$ Point $M = (4, -6)$
    Point $G = (7, -2)$ _____   Point $N = (-1, -6)$ _____

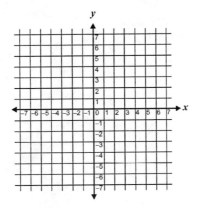

Figure   **Figure**

5. Point $A = (-2, 3)$ 7. Point $J = (-1, 2)$
    Point $B = (-3, 5)$ Point $K = (-1, -1)$
    Point $C = (-1, 6)$ Point $L = (3, -2)$
    Point $D = (1, 5)$
    Point $E = (0, 3)$ _____   _____

6. Point $F = (-1, -3)$ 8. Point $M = (6, 2)$
    Point $G = (-3, -5)$ Point $N = (6, -4)$
    Point $H = (-1, -7)$ Point $O = (4, -4)$
    Point $I = (1, -5)$ _____   Point $P = (4, 2)$ _____

71

Chapter 4 Review

1. What is the distance between the points $(3, 3)$ and $(6, -1)$?

2. What is the distance between the two points $(-3, 0)$ and $(2, 5)$?

For questions 3–6, use the following formula to solve.
$$d = \frac{|am + bn + c|}{\sqrt{a^2 + b^2}}$$

3. What is the distance between the line $y = x + 2$ and the point $(3, 2)$?

4. What is the distance between the line $y = 2x - 1$ and the point $(0, 6)$?

5. What is the distance between the line $y = -4x + 3$ and the point $(5, 0)$?

6. What is the distance between the line $y = \frac{1}{2}x + 1$ and the point $(4, 2)$?

For questions 7 and 8, find the coordinates of the midpoint of the line segments with the given endpoints.

7. $(6, 10)$ $(-4, 4)$

8. $(-1, -7)$ $(5, 3)$

9. Find the missing side.

10. Find the measure of the missing leg of the right triangle below.

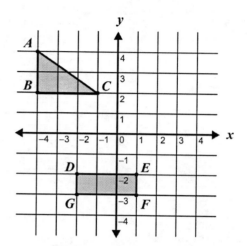

Find the coordinates of the geometric figures graphed above.

11. point A

12. point B

13. point C

14. point D

15. point E

16. point F

17. point G

Plot and label the following points on the same graph.

18. point $H = (1, 1)$

19. point $I = (3, 1)$

20. point $J = (4, -2)$

21. point $K = (2, -2)$

22. What type of figure did you plot?

Chapter 4 Test

1.

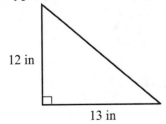

Logan enjoys taking his dog to the park. He walks down Hickory St., turns onto Maple Ave. to meet his friend, Brett, and then continues on Maple Ave. to the park. How far will Brett walk from his house to the park?

(A) 350 yards
(B) 437 yards
(C) 687 yards
(D) 532 yards

2. Approximately what is the measure of the hypotenuse of the triangle?

12 in

13 in

(A) 14 in
(B) 157 in
(C) 18 in
(D) 313 in

3. What is the measure of the missing side in the triangle?

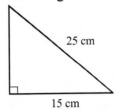

25 cm

15 cm

(A) 5 cm
(B) 29 cm
(C) 10 cm
(D) 20 cm

4. Find the midpoint of a line segment that has the endpoints $(-6, 2)$ and $(0, -2)$.

(A) $(0, -1)$
(B) $(-3, 4)$
(C) $(0, -4)$
(D) $(-3, 0)$

5. Randy plotted points G and H on the Cartesian coordinate graph below. Where could he plot points J and K if he wants to form a square $GHJK$?

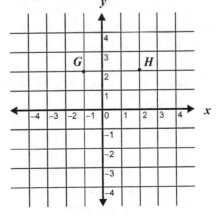

(A) J at $(2, 4)$ and K at $(-1, 4)$
(B) J at $(3, 0)$ and K at $(0, 0)$
(C) J at $(2, 0)$ and K at $(-1, 0)$
(D) J at $(2, -1)$ and K at $(-1, -1)$

6. What is the distance between the line $y = 4x + 1$ and the point $(5, 2)$? Use the following formula:

$$d = \frac{|am + bn + c|}{\sqrt{a^2 + b^2}}$$

(A) 5.00
(B) 4.61
(C) 3.89
(D) 4.17

7. What is the distance between the line $y = -2x + 1$ and the point $(-4, 6)$? Use the following formula:

$$d = \frac{|am + bn + c|}{\sqrt{a^2 + b^2}}$$

(A) 1.34
(B) 2.06
(C) 1.21
(D) 3.74

8. What is the distance between the line $y = x - 5$ and the point $(1, 1)$? Use the following formula:

$$d = \frac{|am + bn + c|}{\sqrt{a^2 + b^2}}$$

(A) 4.02
(B) 3.54
(C) 3.61
(D) 2.98

Chapter 5
Triangles

This chapter covers the following Georgia Performance Standards:

MA1G	Geometry	MA1G3a
		MA1G3b
		MA1G3c
		MA1G3e

5.1   Types of Triangles

A triangle is an **equilateral triangle** if all of its sides and angles are equal. A triangle is an **isosceles triangle** if two of its sides and the angles opposite those sides are equal. A triangle is a **right triangle** if one of its angles equals 90°.

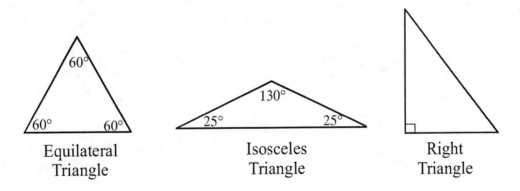

5.2   Interior Angles of a Triangle

The three interior angles of a triangle always add up to 180°.

Example 1:

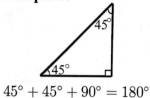

$$45° + 45° + 90° = 180° \qquad 30° + 60° + 90° = 180° \qquad 60° + 60° + 60° = 180°$$

Example 2:   Find the missing angle in the triangle.

Solution:

$$
\begin{aligned}
20° + 125° + x &= 180° \\
-20° \quad -125° & \qquad -20° \; -125° \\
x &= 180° - 20° - 125° \\
x &= 35°
\end{aligned}
$$

Subtract 20° and 125° from both sides to get x by itself.

The missing angle is 35°.

Find the missing angle in the triangles.

1.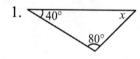

4.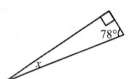

7.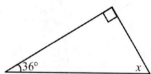

2.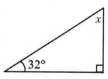

5.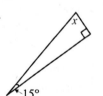

8.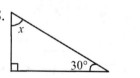

3.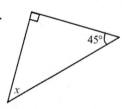

6.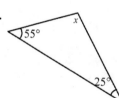

9.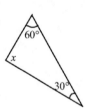

Find the missing angles in the triangles.

10.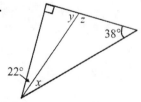

11.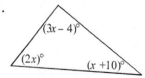

12.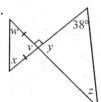

5.3   Exterior Angles

The **exterior angle** of a triangle is always equal to the sum of the opposite interior angles.

Example 3:   Find the measure of $\angle x$ and $\angle y$.

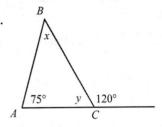

Step 1:   Using the rule for exterior angles,
$120° = \angle A + \angle B$
$120° = 75° + x$
$45° = x$

Step 2:   The sum of the interior angles of a triangle equals $180°$, so
$180° = 75° + 45° + y$
$60° = y$

Find the measures of x and y.

1.
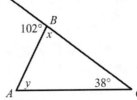

3.
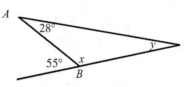

5.
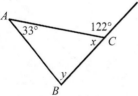

2.
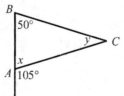

4.
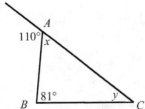

6.
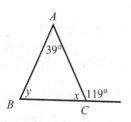

Find the measures of the angles.

7.
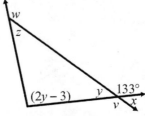

8.
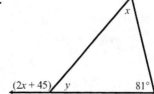

9.
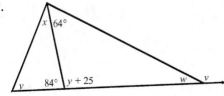

5.4   Side and Angle Relationships

The side-angle inequality theorem states that the longest side must be opposite the largest angle in the triangle. Also, the smallest angle must be opposite the shortest side in the triangle.

Example 4:   Given $\triangle ABC$, list the sides from shortest to longest.

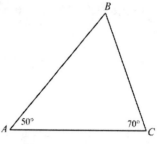

Step 1:   First, we need to find the measure of $\angle B$.
The sum of the measures of the interior angles of a triangle is $180°$, so
$m\angle A + m\angle B + m\angle C = 180°$.
$50° + m\angle B + 70° = 180°$
$m\angle B = 60°$.

Step 2:   By theorem, we know that if one angle of a triangle is larger than another angle, then the side **opposite** the larger angle is longer than the side opposite the smaller angle. Similarly, the side opposite the smallest angle of a triangle is the shortest side.

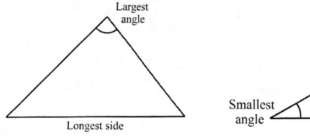

Therefore, in $\triangle ABC$, the shortest side is $\overline{BC}$, which is across from the $50°$ angle. The longest side is $\overline{AB}$, which is across from the $70°$ angle. In order from shortest to longest, then, the sides are $\overline{BC}$, $\overline{AC}$, and $\overline{AB}$. $\overline{BC} < \overline{AC} < \overline{AB}$

Solve the following problems.

1. List the sides of the triangle in order from longest to shortest.

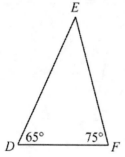

2. List the angles of the triangle in order from smallest to largest.

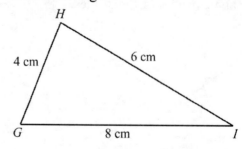

5.5   Triangle Inequality Theorem

The triangle inequality theorem states that the sum of the measure of any two sides in a triangle must be greater than the measure of the third side.

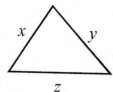

$$x + y > z$$
$$y + z > x$$
$$x + z > y$$

Example 5:   Determine whether or not it is possible to create a triangle with sides of 1 units, 5 units, and 7 units.

Step 1:   First, you must set up three inequalities. Remember the sum of any two sides of a triangle must be greater than the third side.

$$1 + 5 > 7 \qquad 1 + 7 > 5 \qquad 5 + 7 > 1$$

Step 2:   Determine if the inequalities are true.

$$1 + 5 > 7 \qquad 1 + 7 > 5 \qquad 5 + 7 > 1$$
$$6 > 7 \qquad\quad 8 > 5 \qquad\quad 12 > 1$$
$$\text{False} \qquad\quad \text{True} \qquad\quad \text{True}$$

The number 6 is not greater than 7, so a triangle cannot be formed using the sides given.
(All three inequalities must be true in order to create a triangle.)

Determine whether or not it is possible to create a triangle given the following measures of sides. Write yes if it is possible to form a triangle with the given measures of sides or write no if it is not possible.

1. 7, 8, 13

2. 2, 5, 9

3. 10, 8, 15

4. 6, 9, 20

5. 101, 89, 150

6. 1, 2, 4

7. 7, 7, 14

8. 21, 15, 29

9. 11, 9, 17

5.6   Exterior-Angle Inequality

The exterior angle inequality states that an exterior angle of a triangle is equal in measure to the sum of the two non-adjacent interior angles of the triangle.

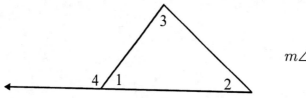

$$m\angle 4 = m\angle 2 + m\angle 3$$

Example 6:   Find the measure of x.

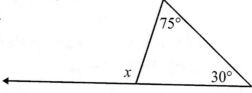

Step 1:   Since you know that x equals the sum of the measures of the two non-adjacent interior angles of the triangle, set up your equation.
$x = 75° + 30°$

Step 2:   Solve for x. $x = 75° + 30° = 105°$

Example 7:   Find the measure of x.

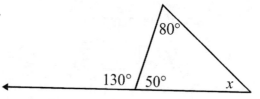

Step 1:   Set up your equation using the exterior angle inequality.
$130° = 80° + x$

Step 2:   Solve for x. $x = 130° - 80° = 50°$

Find the value of x.

1.

3.

5.

2.

4.

6.

Copyright © American Book Company

5.7   Congruence Postulates

Some triangles are congruent and some are similar. There are specific principles that determine whether are not two triangles are congruent or similar. Two triangles are congruent, denoted by the symbol $\cong$, if all of the corresponding sides and angles of each triangle are equal to each other.

Side-Side-Side (SSS) Congruence Postulate

If three sides of one triangle are congruent to three sides of another triangle, then the two triangles are congruent.

If $\overline{AB} \cong \overline{xy}$, and $\overline{BC} \cong \overline{yz}$, and $\overline{AC} \cong \overline{xz}$, then $\triangle ABC \cong \triangle xyz$

Side-Angle-Side (SAS) Congruence Postulate

If two sides and the included angle of one triangle are congruent to two sides and the included angle of the second triangle, then the two triangles are congruent.

If $\overline{AB} \cong \overline{xy}$, and $\angle A \cong \angle x$, and $\overline{AC} \cong \overline{xz}$, then $\triangle ABC \cong \triangle xyz$

Angle-Side-Angle (ASA) Congruence Postulate

If two angles and the included side of one triangle are congruent to two angles and the included side of the second triangle, then the two triangles are congruent.

If $\angle B \cong \angle y$, and $\angle C \cong \angle z$, and $\overline{BC} \cong \overline{yz}$, then $\triangle ABC \cong \triangle xyz$

Angle-Angle-Side (AAS) Congruence Theorem

If two angles and a nonincluded side of one triangle are congruent to two angles and the corresponding side of a second triangle, then the two triangles are congruent.

If $\angle A \cong \angle x$, and $\angle C \cong \angle z$, and $\overline{BC} \cong \overline{yz}$, then $\triangle ABC \cong \triangle xyz$

Hypotenuse-Leg (HL) Theorem

If a hypotenuse and a leg of one right triangle are congruent to a hypotenuse and a leg of another right triangle, then the two triangles are congruent.

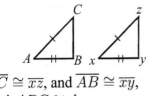

If $\overline{AC} \cong \overline{xz}$, and $\overline{AB} \cong \overline{xy}$, then $\triangle ABC \cong \triangle xyz$

5.8   Points of Concurrency

There are four special points in a triangle called the **centers** of a triangle and they are:

1. **Centroid** - point of concurrency of the medians

2. **Orthocenter** - point of concurrency of the altitudes

3. **Incenter** - point of concurrency of the angle bisectors

4. **Circumcenter** - point of concurrency of the perpendicular bisectors of the sides

Example 8:   The star denotes the **centroid** of $\triangle ABC$.

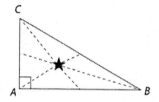

The centroid is the point where the three **medians** of the triangle intersect.

Example 9:   The star denotes the **orthocenter** of $\triangle ABC$.

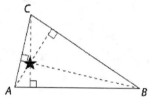

The orthocenter is the point at which the three **altitudes** of the triangle intersect.

Example 10:   The star denotes the **incenter** (the center of a circle inscribed in a triangle) of $\triangle ABC$.

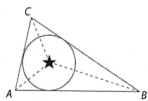

The incenter is located at the point of intersection of the triangle's **angle bisectors**.

Example 11:   The star denotes the **circumcenter** (the center of a circle circumscribed around the triangle).

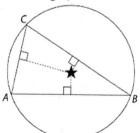

The circumcenter is the point of intersection of the triangle's **perpendicular bisectors**.

Example 12: Below is a chart that lists the type of triangle and the concurrency location.

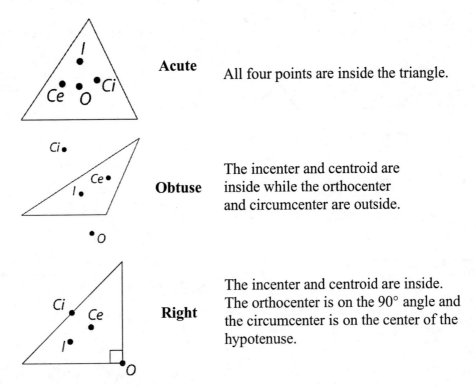

Acute   All four points are inside the triangle.

Obtuse   The incenter and centroid are inside while the orthocenter and circumcenter are outside.

Right   The incenter and centroid are inside. The orthocenter is on the 90° angle and the circumcenter is on the center of the hypotenuse.

Fill in the chart giving the location (in/out/on).

	Centroid	Orthocenter	Incenter	Circumcenter
Acute				
Right				
Obtuse				

True/False.

1. The altitudes meet at the circumcenter.

2. The medians meet at the incenter.

3. The medians meet at the centroid.

4. The perpendicular bisectors of the sides meet at the orthocenter.

Chapter 5 Review

1. Find the missing angle.

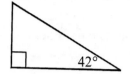

2. What is the largest angle of $\triangle ABC$?

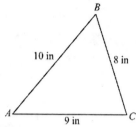

3. Find x.

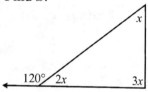

4. What type of triangle has only two equal sides?

5. Find the measures of x and y.

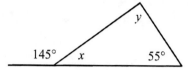

6. Find the measures of x and y.

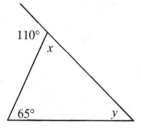

For questions 7–10, determine which inequality theorem is shown: side-angle inequality, triangle inequality, or exterior-angle inequality.

7. $\overline{AB} > \overline{BC}$

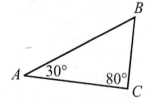

8. $115° = 55° + x$

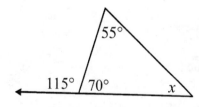

9. $m\angle A > m\angle C$

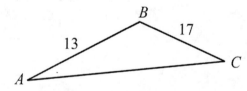

10. $\overline{AB} + \overline{BC} > \overline{AC}$

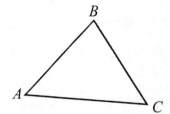

Determine whether the statements about points of concurrency below are true or false.

11. The circumcenter is also the center of the circle inscribed in a triangle.

12. All four points lie within a right triangle.

13. The incenter is where the angle bisectors meet.

14. The orthocenter is always inside the triangle.

15. In a right triangle, the circumcenter is on the middle of the hypotenuse.

16. The centroid is where the medians meet.

17. The centroid is always inside the triangle.

18. The incenter and circumcenter always lie within a triangle.

Determine if the following triangles are congruent. If they are congruent, state the theorem, SSS, SAS, ASA, AAS, or HL, used to prove they are congruent.

19.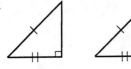

22.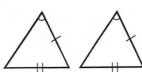

20.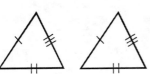

23.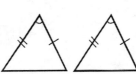

21.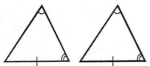

24.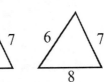

85

Chapter 5 Test

1. What is the measure of missing angle?

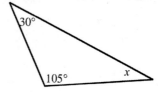

   (A) 225°
   (B) 45°
   (C) 75°
   (D) 30°

2. By what theorem can you prove the two triangles congruent?

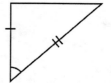

   (A) SAS
   (B) ASA
   (C) SSA
   (D) AAS

3. What is the measure of y?

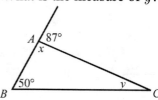

   (A) 93°
   (B) 87°
   (C) 37°
   (D) Cannot be determined

4. Which two points are always inside the triangle?

   (A) Orthocenter & Centroid
   (B) Centroid & Incenter
   (C) Incenter & Orthocenter
   (D) Circumcenter & Centroid

5. What is the measure of the two missing angles?

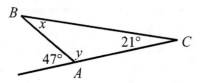

   (A) $x = 26°, y = 133°$
   (B) $x = 47°, y = 112°$
   (C) $x = 112°, y = 47°$
   (D) $x = 133°, y = 26°$

6. What type of triangle is illustrated?

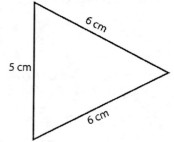

   (A) right
   (B) isosceles
   (C) equilateral
   (D) obtuse

7. Which side in $\triangle ABC$ is the longest side?

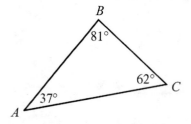

   (A) $\overline{AB}$
   (B) $\overline{AC}$
   (C) $\overline{BC}$
   (D) Cannot be determined

8. The incenter is also the center of the circle inscribed inside the triangle.

(A) True
(B) False - The incenter is outside.
(C) False - The orthocenter is the center of the inscribed circle.
(D) False - The incenter is the center of the circumscribed circle.

9. Which of the following lies on the middle of the hypotenuse?

(A) Incenter
(B) Centroid
(C) Orthocenter
(D) Circumcenter

10. Choose the theorem that proves the congruence.

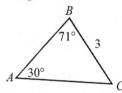 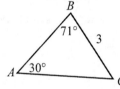

(A) SSS
(B) SAS
(C) AAS
(D) The triangles are not congruent.

11. The meeting place of the altitudes is the

(A) centroid.
(B) incenter.
(C) orthocenter.
(D) circumcenter.

12. The angle bisectors meet at the

(A) centroid.
(B) incenter.
(C) orthocenter.
(D) circumcenter.

13. Which inequality theorem states that the sum of the measures of any two sides in a triangle must be greater than the measure of the third side in that triangle?

(A) side-angle inequality
(B) triangle inequality
(C) exterior-angle inequality
(D) hypotenuse-length theorem

14. Perpendicular bisectors of the sides meet at the

(A) centroid.
(B) incenter.
(C) orthocenter.
(D) circumcenter.

15. Which angle in the triangle below has the smallest measure?

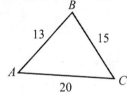

(A) ∠A
(B) ∠B
(C) ∠C
(D) All the angles are the same measure.

16. Choose the theorem that proves the congruence.

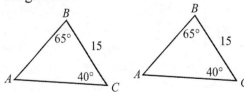

(A) ASA
(B) SAS
(C) AAS
(D) HL

Chapter 6
Quadrilaterals and Other Polygons

This chapter covers the following Georgia Performance Standards:

MA1G	Geometry	MA1G3a
		MA1G3d

6.1 Polygons

A **polygon** is a geometric figure with a finite number of sides in which each side intersects with two other sides only at their endpoints. A **convex polygon** is one in which every line segment connecting any pair of its points is entirely inside the figure's interior. If a polygon is not convex, it is a **concave polygon**. A concave polygon has an indentation and at least one interior angle that is greater than 180°. A **regular polygon** is one in which all the sides are the same measurement and all the angles are the same measurement (congruent).

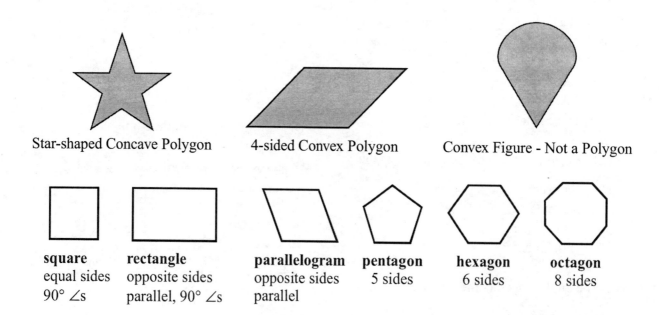

Star-shaped Concave Polygon   4-sided Convex Polygon   Convex Figure - Not a Polygon

square
equal sides
90° ∠s

rectangle
opposite sides
parallel, 90° ∠s

parallelogram
opposite sides
parallel

pentagon
5 sides

hexagon
6 sides

octagon
8 sides

6.2   Sum of Interior Angles of a Polygon

Given a polygon, you can find the sum of the measures of the interior angles using the following formula: Sum of the measures of the interior angles $= 180° (n - 2)$, where n is the number of sides of the polygon.

Example 1:   Find the sum of the measures of the interior angles of the following polygon:

Solution: The figure has 8 sides. Using the formula we have
$180° (8 - 2) = 180° (6) = 1080°$

Using the formula, $180° (n - 2)$, find the sum of the interior angles of the following figures.

1.

4.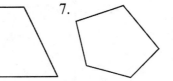

7.

10.

2.

5.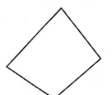

8.

11.

3.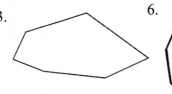

6.

9.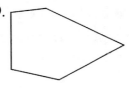

12.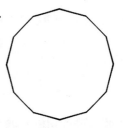

Find the measure of $\angle G$ in the regular polygons shown below. Remember that the sides of a regular polygon are equal.

13.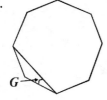

14.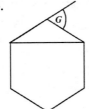

15.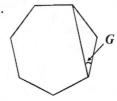

6.3   Exterior Angles of Polygons

The sum of the measures of the exterior angles of a convex polygon, one at each vertex, is 360°.

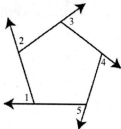

$$m\angle 1 + m\angle 2 + m\angle 3 + m\angle 4 + m\angle 5 = 360°.$$

In addition, for a regular n-gon, in which all angle measures are the same, the measure of each exterior angle is $\dfrac{1}{n} \times 360°$ or $\dfrac{360°}{n}$ where n is the number of sides.

Example 2:   Find the value of x.

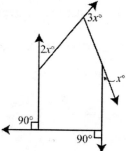

Solution: The sum of the measures of the exterior angles is 360°, so

$$
\begin{aligned}
90° + 90° + x + 2x + 3x &= 360° \\
180° + 6x &= 360° \\
6x &= 180° \\
x &= 30°
\end{aligned}
$$

Example 3:   One exterior angle of a regular convex n-gon measures 22.5°. How many sides does it have?

Solution: The measure of each exterior angle of a convex n-gon is $\dfrac{360°}{n}$, where n is the number of sides. Therefore,

$$22.5° = \frac{360°}{n} \qquad \text{so} \qquad n = \frac{360°}{22.5°} = 16$$

The polygon has 16 sides.

Solve the following problems.

1. Find the measure of each exterior angle of a regular dodecagon (12 sides).

2. Find the measure of each exterior angle of a regular octagon.

3. Each exterior angle of a regular polygon equals 40°. How many sides does the polygon have?

4. The exterior angles of a given polygon measure 45°, 125°, $2x°$, $3x°$, and $5x°$. Find the value of x and list the five angle measures.

5. If the measure of the exterior angles of a polygon are $x°$, $3x°$, $(x + 2)°$, $(5x - 4)°$, 30°, and 42°, find the value of x.

6.4   Quadrilaterals and Their Properties

A **quadrilateral** is a polygon with four sides. A **parallelogram** is a quadrilateral in which both pairs of opposite sides are parallel. The following properties of parallelograms are given without proof:

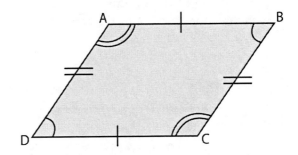

1. **Both pairs of opposite sides are parallel.**

2. **The opposite sides are congruent.**

3. **The opposite angles are congruent**

4. **Consecutive angles are supplementary.**

5. **The diagonals bisect each other**

A **rectangle** is a parallelogram with four right angles. It follows that a rectangle has all of the properties listed above, plus all four angles are 90°. In addition, the diagonals of a rectangle are congruent. A **rhombus** is a parallelogram with four congruent sides. A rhombus has all the properties of a parallelogram, but both pairs of opposite sides are congruent as well as parallel. Due to this fact, the diagonals of a rhombus each bisect a pair of opposite angles and the diagonals are perpendicular to each other. A **square** is a rhombus with four right angles. Therefore, a square has four congruent sides and four congruent angles (each 90°). As you can see, a square is also a quadrilateral, a parallelogram, a rectangle, and a rhombus. These properties plus the four right angles make it a square.

A **trapezoid** is a quadrilateral with only one pair of parallel sides. The parallel sides are called bases, and the other two sides are called legs.

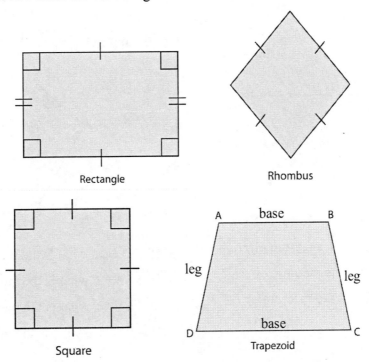

Rectangle

Rhombus

Square

Trapezoid

6.5   More Properties of Quadrilaterals

There are two more kinds of quadrilaterals we need to know about: parallelograms and kites.

Examples:

1. **Parallelogram** – a figure with two sets of parallel sides. The opposite sides are equal in length and the opposite angles are congruent.

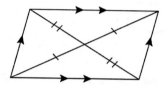

2. **Kite** – a figure with two pairs of congruent adjacent sides. One diagonal divides the kite into two congruent triangles. Two diagonals divide the kite into 4 triangles.

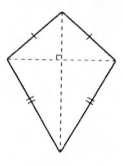

Use the following figure for questions 1–4.

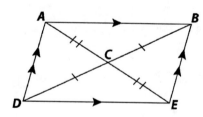

1. If $\overline{AB} = 6$, what is the measure of $\overline{DE}$?

2. If $\overline{AD} = 3$, what is the measure of $\overline{BE}$?

3. If $\overline{AC} = 5$, what is the measure of $\overline{AE}$?

4. If $\overline{BD} = 10$, what is the measure of $\overline{DC}$?

Use the following figure for questions 5–8.

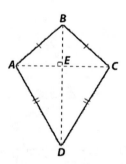

5. If $\overline{CD} = 5$, what is the measure of $\overline{AD}$?

6. If $\overline{AB} = 2$, what is the measure of $\overline{BC}$?

7. If $\overline{AE} = 3$, what is the measure of $\overline{AC}$?

8. If $\overline{CE} = 10$, what is the measure of $\overline{AE}$?

Chapter 6 Review

1. What is the sum of the measures of the interior angles in the figure below?

2. What is the value of one exterior angle of a regular decagon (10 sides)?

3. What is the sum of the exterior angles of a 28-gon?

Use the figure below to answer problems 4–6.

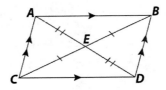

4. True or false. $\triangle ACD \cong \triangle DBA$

5. If $\overline{AC} = 2$, what is the measure of $\overline{BD}$?

6. If $\overline{CE} = 7$, what is the measure of $\overline{CB}$?

Use the figure below to answer problems 7–8.

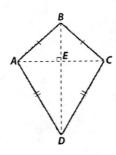

7. True or false. $\triangle ABE \cong \triangle CBE$

8. If $\overline{BC} = 6$, what is the measure of $\overline{AB}$?

Chapter 6 Test

1. What is the sum of the measures of the interior angles of a polygon with 7 sides?

   (A) 180°
   (B) 360°
   (C) 540°
   (D) 900°

2. Which of the following figures is a parallelogram?

   (A)

   (B)

   (C)

   (D)

   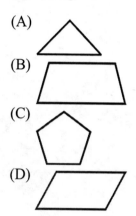

3. Which item below is not a polygon?

   (A) triangle
   (B) heptagon
   (C) octagon
   (D) circle

4. Find the values of x and y in the figure.

   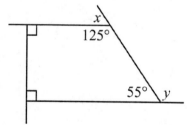

   (A) 55°, 125°
   (B) 90°, 90°
   (C) 55°, 90°
   (D) 90°, 125°

5. If a regular octagon has eight sides, what is the measure of each exterior angle?

   (A) 15°
   (B) 30°
   (C) 45°
   (D) 360°

6. What is the name of the polygon below?

   (A) quadrilateral
   (B) pentagon
   (C) hexagon
   (D) octagon

7. A square is

   (A) a quadrilateral.
   (B) parallelogram.
   (C) both.
   (D) neither.

8. A parallelogram with four congruent sides is

   (A) a rhombus.
   (B) a rectangle.
   (C) a square.
   (D) both A and C.

9. What is the sum of the interior angles of an octagon?

   (A) 1440°
   (B) 1080°
   (C) 1260°
   (D) 360°

10. What is the measure of angle B?

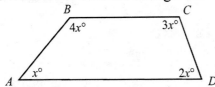

(A) 36°
(B) 90°
(C) 108°
(D) 144°

11. Find the value of x in the figure.

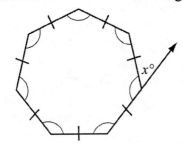

(A) 45°
(B) 51.4°
(C) 90°
(D) 360°

12. Lem wanted to cut a rusted stop sign into quarters in his shop class. Which of the following statements is factual?

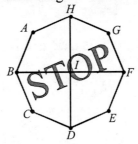

(A) $m\angle ABF = 67.5°$
(B) $m\angle CBF + m\angle BAH = 135°$
(C) $m\angle HGF = 67.5°$
(D) $m\angle BAH - m\angle HIF = 35°$

Chapter 7
Circles and Spheres

This chapter covers the following Georgia Performance Standards:

MA1G	Geometry	MA1G4a
		MA1G4b
		MA1G4c
		MA1G4d
		MA1G5a
		MA1G5b

7.1   Parts of a Circle

A **circle** is defined as all points in a plane that are an equal distance from a point called the **center**. The circle is named by the center point.

A **chord** is a segment that has its endpoints on the circle.

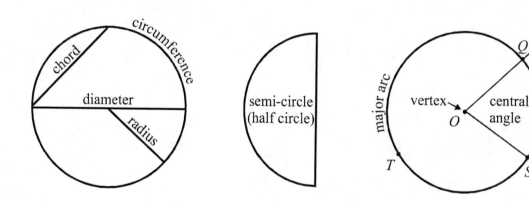

A **central angle** of a circle has the center of the circle as its vertex. The rays of a central angle each contain a radius of the circle. $\angle QOS$ is a central angle.

The points Q and S separate the circle into **arcs**. The arc lies on the circle itself. It does not include any points inside or outside the circle. $\overparen{QRS}$ or $\overparen{QS}$ is a **minor arc** because it is less than a semicircle. A minor arc can be named by 2 or 3 points. $\overparen{QTS}$ is a **major arc** because it is more than a semicircle. A major arc must be named by 3 points.

An **inscribed angle** is an angle whose vertex lies on the circle and whose sides contain **chords** of the circle. $\angle ABC$ in Figure 1 is an inscribed angle.

A line is **tangent** to a circle if it only touches the circle at one point, which is called the point of tangency. See Figure 2 for an example.

A **secant**, shown in Figure 3, is a line that intersects with a circle at two points. Every secant forms a chord. In Figure 3, secant $\overleftrightarrow{AB}$ forms chord $\overline{AB}$.

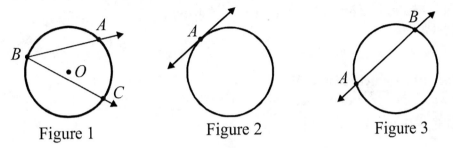

Figure 1   Figure 2   Figure 3

Refer to the figure below, and answer the following questions.

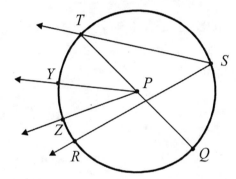

1. Identify the 2 line segments that are chords of the circle but not diameters.

2. Identify the largest major arc of the circle that contains point S.

3. Identify the center of the circle.

4. Identify the inscribed angle(s).

5. Identify the central angle(s).

6. Identify the diameters(s).

7. Identify the secant(s).

8. Name one major arc that has T as one of its ends.

7.2   Arc Lengths

The measure of a minor arc is the measure of its central angle.

In the circle at right, $\angle AOC$ measures 80°.

Therefore, $m\overset{\frown}{AC} = 80$.

A complete rotation about the center point of a circle is 360°.

The measure of a major arc is 360 minus the measure of its central angle.

In the circle at right, $m\overset{\frown}{ADC} = 360 - 80 = 280$.

The measure of a semicircle is 180°.

In the circle at right, $\overline{AD}$ is a diameter of the circle.

Therefore, $\overset{\frown}{ACD}$ is a semicircle and $m\overset{\frown}{ACD} = 180$.

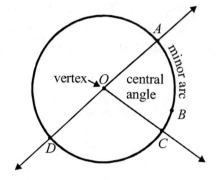

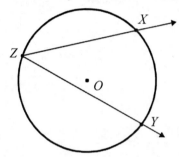

If an angle is inscribed in a circle, then the measure of the minor arc is two times the measure of the inscribed angle.

In the circle at left, $\angle XZY$ measures 45°.

Therefore, $m\overset{\frown}{XY} = 2 \times 45$.

Simplified, $m\overset{\frown}{XY} = 90$.

In the circle below, $m\angle KOJ = 26°$, $m\angle MON = 37°$, and $\overline{KM}$ and $\overline{JL}$ are diameters. Find each measure.

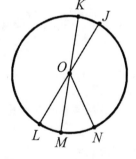

1. $m\overset{\frown}{NM} = $ _____

2. $m\overset{\frown}{KJ} = $ _____

3. $m\overset{\frown}{LM} = $ _____

4. $m\overset{\frown}{JN} = $ _____

5. $m\overset{\frown}{LKJ} = $ _____

6. $m\overset{\frown}{LKN} = $ _____

7. $m\overset{\frown}{MJK} = $ _____

8. $m\overset{\frown}{MNJ} = $ _____

In the circle below, $\overline{AC}$ is a diameter. $\angle DAC = 45°$, $\angle BCD = 110°$, $m\overset{\frown}{AD} = 90$, and $m\overset{\frown}{BC} = 50$. Find each measure.

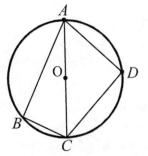

9. $\angle ACD = $ _____

10. $\angle BAC = $ _____

11. $\angle BCA = $ _____

12. $\angle ABC = $ _____

13. $m\overset{\frown}{ABC} = $ _____

14. $\angle CDA = $ _____

15. $m\overset{\frown}{CD} = $ _____

16. $m\overset{\frown}{BCD} = $ _____

7.3   More Circle Properties

An angle created by two secants or by a secant and a tangent has a measure equal to half the difference of the corresponding arc measures.

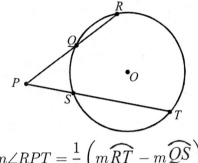

$$m\angle RPT = \frac{1}{2}\left(m\widehat{RT} - m\widehat{QS}\right)$$

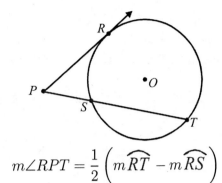

$$m\angle RPT = \frac{1}{2}\left(m\widehat{RT} - m\widehat{RS}\right)$$

If two chords intersect, the angles created have a measure equal to the average of the corresponding arc measures.

Two angles that correspond to the same arc are equal.

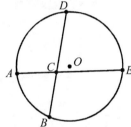

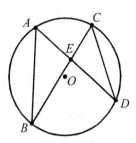

$$m\angle ACB = m\angle DCE = \frac{1}{2}\left(m\widehat{AB} + m\widehat{DE}\right)$$

$$m\angle ACD = m\angle BCE = \frac{1}{2}\left(m\widehat{AD} + m\widehat{BE}\right)$$

$$m\angle ABC = \frac{1}{2}m\widehat{AC}$$

$$m\angle ADC = \frac{1}{2}m\widehat{AC}$$

Use the circles to find the measures of the angles and arcs.

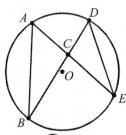

$m\angle BDE = 50°$
$m\angle ABD = 25°$
$m\widehat{AB} = 120$

1. $m\widehat{AD}$
2. $m\angle ACD$
3. $m\angle AED$
4. $m\angle ACB$

5. $m\widehat{DE}$

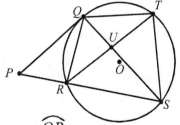

$m\angle QTR = 20°$
$m\angle TSQ = 25°$
$m\angle TRS = 60°$

6. $m\widehat{QR}$
7. $m\angle TUS$
8. $m\angle QPS$

9. $m\widehat{QT}$
10. $m\angle RUS$

7.4   Comparing Diameter, Radius, and Circumference in a Circle

Circumference, C, is the distance around the outside of a circle. ($C = 2\pi r$ or $C = \pi d$)

Diameter, d, is a line segment passing through the center of a circle from one side to the other.

Radius, r, is a line segment from the center of a circle to the edge of the circle.

Pi, π, is the ratio of the circumference of a circle to its diameter. $\left(\pi = 3.14 \text{ or } \frac{22}{7}\right)$

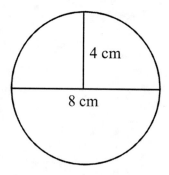

Answer the following questions about the circle above.

1. What is radius of the circle?

2. What is the diameter of the circle?

3. Pi is the ratio of which two parts of a circle?

4. What is the circumference of the circle?

5. What is the symbol for Pi?

6. How many different ways can you draw the radius of the circle?

7. Does the diameter have to pass through the center of the circle?

8. Pi can be written as 3.14. How else can it be written?

7.5   Circumference

Circumference, C - the distance around the outside of a circle
Diameter, d - a line segment passing through the center of a circle from one side to the other
Radius, r - a line segment from the center of a circle to the edge of a circle
Pi, π- the ratio of a circumference of a circle to its diameter $\pi \approx 3.14$ or $\pi \approx \frac{22}{7}$

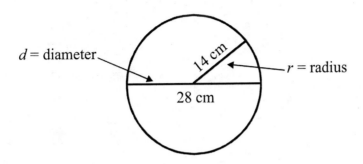

d = diameter   14 cm r = radius
28 cm

The formula for the circumference of a circle is $C = 2\pi r$ or $C = \pi d$. (The formulas are equal because the diameter is equal to twice the radius, $d = 2r$.)

Example 1:   Find the circumference of the circle above.

$$C = \pi d \quad \text{Use } \pi = 3.14$$
$$C = 3.14 \times 28$$
$$C = 87.92 \, \text{cm}$$

$$C = 2\pi r$$
$$C = 2 \times 3.14 \times 14$$
$$C = 87.92 \, \text{cm}$$

Use the formulas given above to find the circumferences of the following circles. Use $\pi = 3.14$.

1. 8 in   2. 14 ft   3. 2 cm   4. 6 m   5. 8 ft

$C = \underline{\quad}$ $C = \underline{\quad}$ $C = \underline{\quad}$ $C = \underline{\quad}$ $C = \underline{\quad}$

Use the formulas given above to find the circumferences of the following circles. Use $\pi = \frac{22}{7}$.

6. 3 ft   7. 12 in   8. 6 m   9. 5 cm   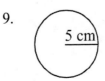   10. 16 in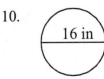

$C = \underline{\quad}$ $C = \underline{\quad}$ $C = \underline{\quad}$ $C = \underline{\quad}$ $C = \underline{\quad}$

7.6   Area of a Circle

The formula for the area of a circle is $A = \pi r^2$. The area is how many square units of measure would fit inside a circle.

Example 2:   Find the area of the circle, using both values for π.

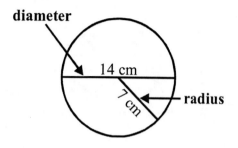

diameter

14 cm

7 cm

radius

Let $\pi = \frac{22}{7}$
$A = \pi r^2$
$A = \frac{22}{7} \times 7^2$
$A = \dfrac{22}{\overset{}{7}\ 1} \times \dfrac{\overset{49}{\cancel{49}}\ 7}{1}$

$= 154$ cm^2

Let $\pi = 3.14$
$A = \pi r^2$
$A = 3.14 \times 7^2$
$A = 3.14 \times 49$

$= 153.86$ cm^2

Find the area of the following circles. Remember to include units.

		$\pi = 3.14$	$\pi = \frac{22}{7}$
1.	5 in	$A =$ _____	$A =$ _____
2.	16 ft	$A =$ _____	$A =$ _____
3.	8 cm	$A =$ _____	$A =$ _____
4.	3 m	$A =$ _____	$A =$ _____

Fill in the chart below. Include appropriate units.

			Area	
	Radius	Diameter	$\pi = 3.14$	$\pi = \frac{22}{7}$
5.	9 ft			
6.		4 in		
7.	8 cm			
8.		20 ft		
9.	14 m			
10.		18 cm		
11.	12 ft			
12.		6 in		

7.7 Area of Sectors

A sector of a circle is a region bounded by a central angle and its intercepted arc. In the circle below $\angle AOB$ is a central angle measuring $80°$. Therefore, $\overset{\frown}{AB}$ is $80°$. $\angle AOB$ and $\overset{\frown}{AB}$ form a sector of the circle.

Example 3: Find the area of the sector formed by $\angle AOB$ and $\overset{\frown}{AB}$.

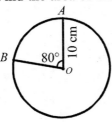

Step 1: The area of a sector is a fraction of the area of the circle. So we must find the area of the circle.
$$A = \pi r^2$$
$$A = 3.14 \times 10^2 = 3.14 \times 100 = 314 \text{ cm}^2$$

Step 2: Now we need to find the fraction of the circle that the sector occupies. Remember that the sum of the measures of the central angles of a circle is 360. The fraction that the sector occupies is the measure of the central angle, denoted by the letter N, divided by 360.
$$\text{Fraction that sector occupies} = \frac{N}{360} = \frac{80}{360} = \frac{2}{9}$$

Step 3: Now we can calculate the area of the sector.
$$A = \frac{N}{360}\pi r^2 = \frac{2}{9} \times 3.14 \text{ cm}^2 \times 10^2 = \frac{628}{9} \text{ cm}^2$$
Simplified, $A = 69\frac{7}{9} \text{ cm}^2$

Each of the following is a measurement for a central angle. Calculate the fraction of a circle that the central angle occupies. Simplify your answers.

1. 30°   3. 16°   5. 45°   7. 120°   9. 15°   11. 60°

2. 2°   4. 52°   6. 72°   8. 270°   10. 108°   12. 90°

Find the area of the sector bounded by $\angle XYZ$ and $\overset{\frown}{XZ}$ in each of the following circles. Use $\pi = 3.14$.

13.
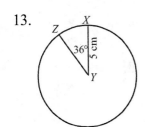

14.
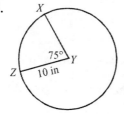

15.
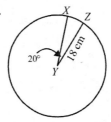

16.
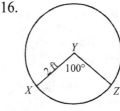

7.8   Volume of Spheres

To find the volume of a solid, insert the measurements given for the solid into the correct formula and solve. Remember, volumes are expressed in cubic units such as in^3, ft^3, m^3, cm^3, or mm^3. The formula for the volume of a sphere is $V = \frac{4}{3}\pi r^3$.

Example 4:   Find the volume of the sphere below. Use $\pi = 3.14$.

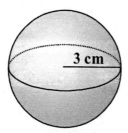

3 cm

Step 1:   Substitute all known values into the equation $V = \frac{4}{3}\pi r^3$.

$r = 3$ $\pi = 3.14$

$V = \frac{4}{3}\pi r^3 = \frac{4}{3}(3.14)(3)^3$

Step 2:   Solve the equation.

$V = \frac{4}{3}(3.14)(3)^3 = \frac{4}{3} \times 3.14 \times 27 = 113.04 \text{ cm}^3$

Find the volume of the following shapes. Use $\pi = 3.14$.

1.
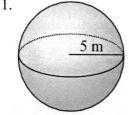
5 m

3.
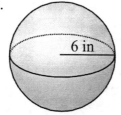
6 in

5.

16 cm

2.
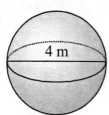
4 m

4.
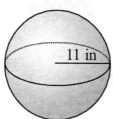
11 in

6.
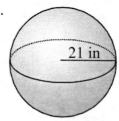
21 in

7. If a basketball measures 24 centimeters in diameter, what volume of air will it hold? Use $\pi = 3.14$.

8. If a vinyl ball measures 5 inches in diameter, what volume of air will it hold? Use $\pi = 3.14$.

9. A spherical dog toy has a radius of 6 cm. If this toy is stuffed with dog treats, what volume of treats will it hold? Use $\pi = 3.14$.

7.9   Surface Area of a Sphere

The formula for the surface area of a sphere is $SA = 4\pi r^2$.

Example 5:   Find the surface area of the sphere below. Use $\pi = 3.14$.

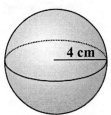

4 cm

Step 1:   Substitute all known values into the equation $SA = 4\pi r^2$.

$r = 4 \quad \pi = 3.14$

$SA = 4\pi r^2 = 4\,(3.14)\,(4)^2$

Step 2:   Solve the equation.

$SA = 4\,(3.14)\,(4)^2 = 4 \times 3.14 \times 16 = 200.96 \text{ cm}^2$

Find the surface area of a sphere given the following measurements where r = radius and d = diameter. Use $\pi = 3.14$.

1. $r = 2$ in SA =   _____   7. $d = 14$ cm SA =   _____

2. $r = 6$ m SA =   _____   8. $r = \frac{1}{5}$ km SA =   _____

3. $r = \frac{3}{4}$ yd SA =   _____   9. $d = 3$ in SA =   _____

4. $d = 8$ cm SA =   _____   10. $d = \frac{2}{3}$ ft SA =   _____

5. $d = 50$ mm SA =   _____   11. $r = 10$ mm SA =   _____

6. $r = \frac{1}{4}$ ft SA =   _____   12. $d = 5$ yd SA =   _____

7.10   Geometric Relationships of Circles and Spheres

This section illustrates what happens to the area or volume of a figure when one or more of the dimensions changes.

Example 6:   How would doubling the radius of a sphere affect the volume?

The volume of a sphere is $V = \frac{4}{3}\pi r^3$. Just by looking at the formula, can you see that by doubling the radius, the volume would increase 8 times the original volume? So, a sphere with a radius of 2 would have a volume 8 times greater than a sphere with a radius of 1.

Example 7:   Sonya drew a circle which had a radius of 3 inches for a school project. She also needed to make a larger circle which had a radius of 9 inches. When Sonya drew the bigger circle, what was the difference in area?

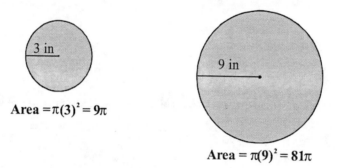

Area $= \pi(3)^2 = 9\pi$

Area $= \pi(9)^2 = 81\pi$

The area of the second circle is 9 times larger than the first.

Carefully read each problem below and solve.

1. Ken draws a circle with a radius of 5 cm. He then draws a circle with a radius of 10 cm. How many times larger is the area of the second circle?

2. The area of circle B is 9 times larger than the area of circle A. If the radius of circle A is represented by x, how would you represent the radius of circle B?

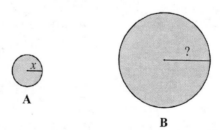

A

B

3. A sphere has a radius of 1. If the radius is increased to 3, how many times greater will the volume be?

4. In a sphere, how many times greater is the volume if you double the diameter?

5. In a sphere, how many times greater is the surface area if you double the radius?

Chapter 7 Review

1. In the circle below, $\overline{AE}$ is a diameter, $\angle DAE$ measures $30°$ and $m\widehat{BC} = 45$. What is the measure of $\widehat{DE}$ and $\angle BOC$?

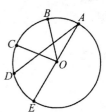

2. Find the area of the shaded part of the image below. Round your answer to the nearest whole number.

3. Calculate the circumference and the area of the following circle. Use $\pi = \frac{22}{7}$.

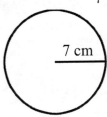

7 cm

4. Calculate the circumference and the area of the following circle. Use $\pi = 3.14$.

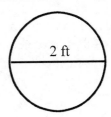

2 ft

5. Which line represents a tangent of the circle?

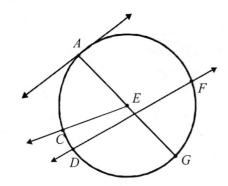

6. Use $\pi = 3.14$.

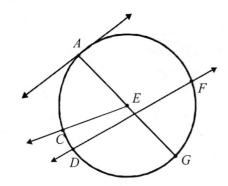

$V = \underline{\hspace{1cm}}$ $SA = \underline{\hspace{1cm}}$

7. If a ball is 4 inches in diameter, what is its surface area? Use $\pi = 3.14$.

8. A gigantic bronze sphere is being added to the top of a tall building downtown. The sphere will be 24 ft in diameter. What will be the surface area of the globe?

9. If a basketball measures 15 inches in diameter, what volume of air will it hold? Use $\pi = 3.14$.

10. In a sphere, how many times greater is the volume if you quadruple the diameter?

Chapter 7 Test

1. What is the area of a circle with a radius of 7 cm? (Round to the nearest whole number)

   (A) 154 square cm
   (B) 196 square cm
   (C) 347 square cm
   (D) 616 square cm

2. Find the circumference. Use $\pi = 3.14$.

   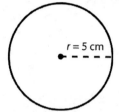

   (A) 15.7 cm
   (B) 62.8 cm
   (C) 31.4 cm
   (D) 0.314 cm

3. Find the area. Use $\pi = 3.14$.

   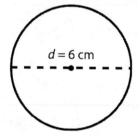

   (A) 113.04 cm²
   (B) 28.26 cm²
   (C) 18.84 cm²
   (D) 188.4 cm²

4. Which is an angle whose vertex lies on a circle and whose sides contain chords of the circle?

   (A) central angle
   (B) tangent
   (C) inscribed angle
   (D) secant

5. A line that touches a circle only at one point is called a

   (A) tangent.
   (B) chord.
   (C) secant.
   (D) inscribed line.

6. What does the measure of $\overset{\frown}{ABC}$ equal?

   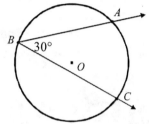

   (A) 60
   (B) 330
   (C) 300
   (D) 210

7. What is the circumference of a circle that has a diameter of 10 cm?

   (A) 15.7 cm
   (B) 31.4 cm
   (C) 78.5 cm
   (D) 310 cm

8. If you decide to divide a pie that has a diameter of 8 in into 6 equal slices, what is the area of each slice?

   (A) 8.37 in²
   (B) 4.19 in²
   (C) 33.49 in²
   (D) 50.24 in²

9. If a sphere with a 6 m radius is cut out of a cube like the one shown below, what would the new volume be? Use $\pi = 3.14$.

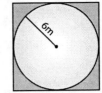

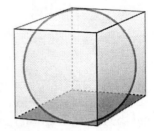

(A) 823.68 m^3
(B) 1728 m^3
(C) 904.78 m^3
(D) 1441 m^3

10. If the radius of a sphere is tripled, how much larger will the volume be?

(A) 81 times larger
(B) 9 times larger
(C) 3 times larger
(D) 27 times larger

11. What is the surface area of a sphere that has a radius that measures 3 feet? Use $\pi = 3.14$.

(A) 113.04 ft^2
(B) 84.82 ft^2
(C) 150.80 ft^2
(D) 201.06 ft^2

12. Find the area of the shaded region.

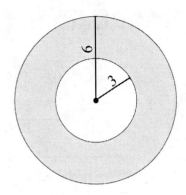

(A) 24π
(B) 3π
(C) 27π
(D) 15π

13. What is the surface area of a sphere whose radius measures 8 cm?

(A) 512 cm^2
(B) 804 cm^2
(C) 384 cm^2
(D) 268 cm^2

14. If you double the diameter of a sphere, how does this effect the surface area.

(A) The surface area becomes four times bigger.
(B) The surface area becomes four times smaller.
(C) The surface area becomes two times bigger.
(D) The surface area becomes two times smaller.

Chapter 8
Permutations and Combinations

This chapter covers the following Georgia Performance Standards:

| MA1D | Data Analysis and Probability | MA1D1a |
| | | MA1D1b |

8.1   Principles of Counting

Principles of counting provide us with counting techniques that can be used to handle large amounts of data. There are two types of principles of counting, the addition rule and the multiplication rule.

Addition Rule: Let X_1 and X_2 be mutually exclusive events (there are no common outcomes). Let X describe the situation where either event X_1 or X_2 will occur. The number of times X may occur is given by:

$n(X) = n(X_1) + n(X_2)$

$n(X) =$ number of outcomes of event X

$n(X_1) =$ number of outcomes of event X_1

$n(X_2) =$ number of outcomes of event X_2

Example 1:   Consider the set $C = \{-7, -4, -1, 3, 5, 7, 9, 11\}$. How many different ways could you choose a number that is positive or even? Find $n(X)$.

Step 1:   Define your variables.
$X =$ choosing a positive or even number from C
$X_1 =$ pick a positive number from C
$X_2 =$ pick an even number from C

Step 2:   Find X_1 and X_2.
$X_1 =$ the positive integers in the set $= \{3, 5, 7, 9, 11\} \rightarrow n(X_1) = 5$
$X_2 =$ the even integers in the set $= \{-4\} \rightarrow n(X_2) = 1$

Step 3:   Substitute the values of X_1 and X_2 into the equation $n(X) = n(X_1) + n(X_2)$.
$n(X) = 5 + 1 = 6$

Example 2:   How many different ways could you choose a number from 1 to 25 that is a multiple of 4 or 7?

Step 1:   Define your variables.

X = choosing a number from 1 to 25 that is either a multiple of 4 or 7

X_1 = the numbers from 1 to 25 that are multiples of 4

X_2 = the numbers from 1 to 25 that are multiples of 7

Step 2:   Find X_1 and X_2.

X_1 = multiples of 4 = $\{4, 8, 12, 16, 20, 24\} \rightarrow n(X_1) = 6$

X_2 = multiples of 7 = $\{7, 14, 21\} \rightarrow n(X_2) = 3$

Step 3:   Substitute the values of X_1 and X_2 into the equation $n(X) = n(X_1) + n(X_2)$.

$n(X) = 6 + 3 = 9$

There are nine ways to choose a number from 1 to 25 that is either a multiple of 4 or 7.

Example 3:   How many different ways could you choose a number from 1 to 10 so that it is a multiple of 2 or 6?

Step 1:   Find X_1 and X_2.

X_1 = multiples of 2 = $\{2, 4, 6, 8, 10\} \rightarrow n(X_1) = 5$

X_2 = multiples of 6 = $\{6\} \rightarrow n(X_2) = 1$

As you can see from the listed multiples, one of the values is the same, 6. Therefore, this problem is not mutually exclusive, so you cannot use the equation for the addition rule. You must use the equation $n(X) = n(X_1) + n(X_2) - n(X_1 \cap X_2)$.

Step 2:   Substitute the values of X_1 and X_2 into the equation $n(X) = n(X_1) + n(X_2) - n(X_1 \cap X_2)$.

$X_1 \cap X_2$ = the values that are multiples of 2 and of 6 = $\{6\} \rightarrow n(X_1 \cap X_2) = 1$

$n(X) = 5 + 1 - 1 = 5$

There are five ways to choose a number from 1 to 10 that is either a multiple of 2 or 6.

Multiplication Rule: Let X_1 and X_2 be independent events (neither will affect the other's outcome) given by: $n(X) = n(X_1) \times n(X_2)$.

Example 4:   Samuel goes to a fast food restaurant for lunch. There are 5 sandwiches and 4 sides to choose from. He wants to pick one sandwich and one side. How many different meals can he order?

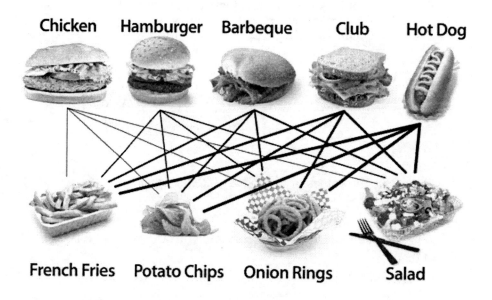

Solution: There are 5 sandwiches and 4 sides to choose from. To find the number of different meals he can get, you must multiply the choices together.
$5 \times 4 = 20$
There 20 different choices Samuel has for lunch.

Example 5:   Stephanie needs to choose a password for her e-mail account. It must be a eight characters long. Four of the characters must be numbers and four of the characters must be letters. How many different password choices does she have?

Step 1:   There are 10 numbers, so there are 10 choices for the first four characters of the password. There are 26 letters, so there are 26 choices for the next four characters of the password.

Step 2:   Now, multiply all the possible choices together to get the total number of choices for the password.
$10 \times 10 \times 10 \times 10 \times 26 \times 26 \times 26 \times 26 = 4,569,760,000$
There are $4,569,760,000$ different possibilities for an eight character password with four numbers and four letters.

Answer the following counting problems.

1. Consider the set $A = \{-7, -6, 1, 3, 5, 7\}$. How many different ways can you pick a positive or even number?

2. Consider the set $A = \{-7, -5, 1, 3, 4, 6, 8\}$. How many different ways can you pick a negative or even number?

3. How many different ways could you choose a number from 1 to 30 so that it is a multiple of 5 or 8?

4. How many different ways could you choose a number from 1 to 10 so that it is a multiple of 4 or 7?

5. Raymond has 7 baseball caps, 2 jackets, 10 pairs of jeans, and 2 pairs of sneakers. How many combinations of the 4 items can he make?

6. Claire has 6 kinds of lipstick, 4 eye shadows, 2 kinds of lip liner, and 2 mascaras. How many combinations can she use to make up her face?

7. Clarence's dad is ordering a new truck. He has a choice of 5 exterior colors, 3 interior colors, 2 kinds of seats, and 3 sound systems. How many combinations does he have to pick from?

8. A fast food restaurant has 8 kinds of sandwiches, 3 kinds of French fries, and 5 kinds of soft drinks. How many combinations of meals could you order if you ordered a sandwich, fries, and a drink?

9. In summer camp, Tyrone can choose from 4 outdoor activities, 3 indoor activities, and 3 water sports. He has to choose one of each. How many combinations of activities can he choose?

10. Jackie won a contest at school and gets to choose one pencil and one pen from the school store and an ice cream from the lunch room. There are 5 colors of pencils, 3 colors of pens, and 4 kinds of ice cream. How many combinations of prize packages can she choose?

8.2   Permutations

A **permutation** is an arrangement of items in a specific order. The formula $_nP_r = \dfrac{n!}{(n-r)!}$ is the formula for permutations. The number n is the number you have to choose from, and r is the number of objects you want to arrange. If a problem asks how many ways can you arrange 6 books on a bookshelf, it is asking how many permutations there are for 6 items.

Example 6:   Ron has 4 items: a model airplane, a trophy, an autographed football, and a toy sports car. How many ways can he arrange the 4 items on a shelf?

Solution: Ron has 4 choices for the first item on the shelf. He then has 3 choices left for the second item. After choosing the second item, he has 2 choices left for the third item and only one choice for the last item. The diagram below shows the permutations for arranging the 4 items on a shelf if he chose to put the trophy first.

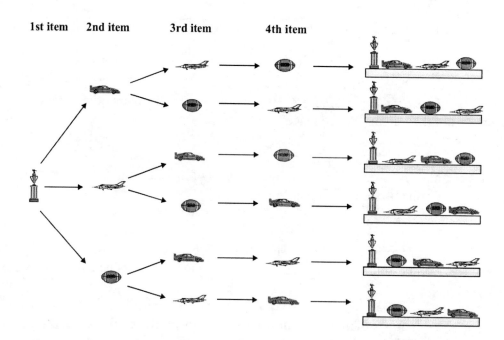

Count the number of permutations if Ron chooses the trophy as the first item. There are 6 permutations. Next, you could construct a pyramid of permutations choosing the model car first. That pyramid would also have 6 permutations. Then, you could construct a pyramid choosing the airplane first. Finally, you could construct a pyramid choosing the football first. You would then have a total of 4 pyramids each having 6 permutations. The total number of permutations is $6 \times 4 = 24$. There are 24 ways to arrange the 4 items on a bookshelf.

You probably don't want to draw pyramids for every permutation problem. What if you want to know the permutations for arranging 30 objects? Fortunately, mathematicians have come up with a formula for calculating permutations.

For the above problem, Ron has 4 items to arrange. Therefore, multiply $4 \times 3 \times 2 \times 1 = 24$. Another way of expressing this calculation is 4!, stated as 4 factorial. $4! = 4 \times 3 \times 2 \times 1$.

Note: To find the permutation using the formula, $_nP_r = \dfrac{n!}{(n-r)!}$. n is the number you have to choose from, 4, and r is the number of objects you want to arrange, 4.

$$_4P_4 = \frac{4!}{(4-4)!} = \frac{4!}{0!} = \frac{4!}{1} = 4! = 4 \times 3 \times 2 \times 1 = 24$$

(Remember that $0! = 1$.)

Example 7:   How many ways can you line up 6 students?

Solution: The number of permutations for 6 students $= 6! = 6 \times 5 \times 4 \times 3 \times 2 \times 1 = 720$. There are 6 choices for the first position, 5 for the second position, 4 for the third, 3 for the fourth, 2 for the fifth, and 1 for the sixth.

Note: The numbers of permutation for 6 students $= {}_6P_6 = \dfrac{6!}{(6-6)!} = \dfrac{6!}{0!} = \dfrac{6!}{1} = 6! = 720$.

Example 8:   Shelley and her mom, dad, and brother are having cake for her birthday. Since it is Shelley's birthday she gets a piece first. How many ways are there to pass out the pieces of cake?

Solution: Since Shelley gets the first piece, the first spot is fixed. The second, third, and fourth spots are not fixed and anyone left can be in one of the three spots.

Spot	1	2	3	4
Choices of people	1	3	2	1

Now, multiply the choices together, $1 \times 3 \times 2 \times 1 = 6$ ways to pass out cake.

Note: You can also use the permutation formula, but fix the first spot and just arrange the last three spots.

$$1 \times {}_3P_3 = \frac{3!}{(3-3)!} = \frac{3!}{0!} = \frac{3!}{1} = 3! = 3 \times 2 \times 1 = 6.$$

Work the following permutation problems.

1. How many ways can you arrange 5 books on a bookshelf?

2. Myra has 6 novels to arrange on a book shelf. How many ways can she arrange the novels?

3. Seven sprinters signed up for the 100 meter dash. How many ways can the seven sprinters line up on the starting line?

4. Keri wants an ice cream cone with one scoop of chocolate, one scoop of vanilla, and one scoop of strawberry. How many ways can the scoops be arranged on the cone?

5. How many ways can you arrange the letters A, B, C, and D?

6. At Sam's party, the DJ has four song requests. In how many different orders can he play the 4 songs?

7. Yvette has 5 comic books. How many different ways can she stack the comic books?

8. Sandra's couch can hold three people. How many ways can she and her two friends sit on the couch?

9. How many ways can you arrange the numbers 2, 3, 5?

10. At a busy family restaurant, four tables open up at the same time. How many different ways can the hostess seat the next four families waiting to be seated?

11. How many ways can you arrange the numbers 1, 2, 3, 4, 5, 6, 7, 8, 9, 10 and always have 3 at position 1 and 10 at position 5?

8.3   More Permutations

The formula $_nP_r = \dfrac{n!}{(n-r)!}$ can also be used if you are trying to arrange a specific number of objects, but have more than you want to arrange. The number n is the number you have to choose from, and r is the number of objects you want to arrange.

Example 9:   If there are 6 students, how many ways can you line up any 4 of them?

  Step 1:   Find all the variables. The formula is $_nP_r = \dfrac{n!}{(n-r)!}$.
  n (the number you have to choose from) = 6
  r (the number of objects you want to arrange) = 4

  Step 2:   Plug the variables into the formula.
  $$_6P_4 = \frac{6!}{(6-4)!} = \frac{6!}{2!} = \frac{6 \times 5 \times 4 \times 3 \times 2 \times 1}{2 \times 1} = 6 \times 5 \times 4 \times 3 = 360$$
  There are 360 ways to line up 4 of the 6 students.

Find the number of permutations for each of the problems below.

1. How many ways can you arrange 4 out of 8 books on a shelf?

2. How many 3 digit numbers can be made using the numbers 2, 3, 5, 8, and 9?

3. How many ways can you line up 4 students out of a class of 20?

4. Kim worked in the linen department of a store. Eight new colors of towels came in. Her job was to line up the new towels on a long shelf. How many ways can she arrange the 8 colors?

5. Terry's CD player holds 5 CDs. Terry owns 12 CDs. How many different ways can he arrange his CDs in the CD player?

6. Erik has 11 shirts he wears to school. How many ways can he choose a different shirt to wear on Monday, Tuesday, Wednesday, Thursday, and Friday?

7. Deb has a box of 12 markers. The art teacher told her to choose three markers and line them up on her desk. How many ways can she line up 3 markers from the 12?

8. Jeff went into an ice cream store serving 32 flavors of ice cream. He wanted a cone with two different flavors. How many ways could he order 2 scoops of ice cream, one on top of the other?

9. In how many ways can you arrange any 3 letters from the 26 letters in the alphabet?

8.4   Combinations

In a **permutation**, objects are arranged in a particular order. In a **combination**, the order does not matter. In a **permutation**, if someone picked two letters of the alphabet, **k, m** and **m, k** would be considered 2 different permutations. In a **combination**, **k, m** and **m, k** would be the same combination. A different order does not make a new combination. The formula for combinations is $_nC_r = \binom{n}{r} = \dfrac{n!}{(n-r)!r!}$ where n is the total number of objects you have to choose from and r is the number that you choose to arrange.

Example 10:   How many combinations of 3 letters from the set {a, b, c, d, e} are there?

    Step 1:   Find the **permutation** of 3 out of 5 objects.

    Step 2:   Divide by the permutation of the **number of objects** to be chosen from the total (3). This step eliminates the duplicates in finding the permutations.

$$\dfrac{5 \times 4 \times 3}{3 \times 2 \times 1} = 10$$

    Step 3:   Cancel common factors and simplify.

Note: Using the formula, find all your variables. The formula is $_nC_r = \binom{n}{r} = \dfrac{n!}{(n-r)!r!}$.

n (the number you choose from) = 5, r (the number of objects arranged) = 3.

$$_5C_3 = \binom{5}{3} = \frac{5!}{(5-3)!3!} = \frac{5!}{2!3!} = \frac{5 \times 4 \times 3 \times 2 \times 1}{(2 \times 1)(3 \times 2 \times 1)} = \frac{20}{2} = 10$$

There can be 10 combinations of three letters from the set {a, b, c, d, e}.

Find the number of combinations for each problem below.

1. How many combinations of 4 numbers can be made from the set of numbers {2, 4, 6, 7, 8, 9}?

2. Johnston Middle School wants to choose 3 students at random from the 7th grade to take an opinion poll. There are 124 seventh graders in the school. How many different groups of 3 students could be chosen? (Use a calculator for this one.)

3. How many combinations of 3 students can be made from a class of 20?

4. Fashion Ware catalog has a sweater that comes in 8 colors. How many combinations of 2 different colors does a shopper have to choose from?

5. Angelo's Pizza offers 10 different pizza toppings. How many different combinations can be made of pizzas with four toppings?

6. How many different combinations of 5 flavors of jelly beans can you make from a store that sells 25 different flavors of jelly beans?

7. The track team is running the relay race in a competition this Saturday. There are 14 members of the track team. The relay race requires 4 runners. How many combinations of 4 runners can be formed from the track team?

8. Kerri got to pick 2 prizes from a grab bag containing 12 prizes. How many combinations of 2 prizes are possible?

Chapter 8 Review

Answer the following problems.

1. Daniel has 7 trophies he has won playing soccer. How many different ways can he arrange them in a row on his bookshelf?

2. Missy has 12 colors of nail polish. She wears 1 color each day, 7 different colors a week. How many combinations of 7 colors can she make before she has to repeat the same 7 colors in a week?

3. Eileen has a collection of 12 antique hats. She plans to donate 5 of the hats to a museum. How many combinations of hat are possible for her donation?

4. Julia has 5 porcelain dolls. How many ways can she arrange 3 of the dolls on a display shelf?

5. Ms. Randal has 10 students. Every day she randomly draws names of 2 students out of a bag to turn in their homework for a test grade. How many combinations of 2 students can she draw?

6. In the lunch line, students can choose 1 out of 3 meats, 1 out of 4 vegetables, 1 out of 3 desserts, and 1 out of 5 drinks. How many lunch combinations are there?

7. Andrea has 7 teddy bears in a row on a shelf in her room. How many ways can she arrange the bears in a row on her shelf?

8. Adrianna has 4 hats, 8 shirts, and 9 pairs of pants. Choosing one of each, how many different clothes combinations can she make?

9. The buffet line offers 5 kinds of meat, 3 different salads, a choice of 4 desserts, and 5 different drinks. If you choose one food from each category, from how many combinations would you have to choose?

10. How many pairs of students can Mrs. Smith choose to go to the library if she has 20 students in her class?

Answer the following counting problems.

11. Consider the set $B = \{-100, -56, -32, 1, 39, 85, 213\}$. How many ways are there to choose a negative or odd number?

12. Consider the set $B = \{-100, -56, -32, 1, 39, 85, 213\}$. How many ways are there to choose a positive or even number?

13. How many different ways could you choose a number from 1 to 30 so that it is a multiple of 4 or 9?

14. How many different ways could you choose a number from 1 to 50 so that it is a multiple of 4 or 6?

15. How many different outfits can Joe make if he has 10 shirts, 6 pairs of pants, and two pairs of shoes?

16. How many different outfits can you make if you have 8 shirts, 4 pairs of pants, and two pairs of shoes?

17. The buffet line offers 5 kinds of meat, 3 different salads, a choice of 4 desserts, and 5 different drinks. If you choose one food from each category, from how many combinations would you have to choose?

Chapter 8 Test

1. How many ways can you arrange six books on a bookshelf?

   (A) 6
   (B) 12
   (C) 36
   (D) 720

2. Katie and Kayla are having a party. They each invite four friends. How many ways could they set around a rectangular table if Katie is always sitting at the left end of the table and Kayla is always at the right end of the table?

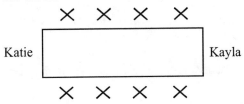

   (A) 80
   (B) 6400
   (C) 40, 320
   (D) 50, 000

3. How many ways can you line up any four students out of a class of twenty?

   (A) 116, 280
   (B) 118, 260
   (C) 260, 118
   (D) 112, 280

4. How many combinations of 3 numbers can be made from the set of numbers $\{2, 3, 5, 6, 7, 9\}$?

   (A) 5
   (B) 10
   (C) 15
   (D) 20

5. Savannah is going to the school dance. She has three dresses, six necklaces, and eight pairs of earrings. How many different outfits does she have to chose from if she chooses only one dress, one necklace, and one pair of earrings?

   (A) 17
   (B) 144
   (C) 48
   (D) 12

Use number set A to answer questions 6–7.

$$A = \{-11, -7, -3, -1, 0, 2, 4, 8, 10, 16\}$$

6. How many ways can you choose a negative or even number from A?

   (A) 11
   (B) 10
   (C) 9
   (D) 7

7. How many ways can you choose a negative or a multiple of 8 number from A?

   (A) 6
   (B) 2
   (C) 4
   (D) 8

8. How many different ways could you choose a number from 1 to 40 so that it is a multiple of 15 or 10?

   (A) 8
   (B) 6
   (C) 5
   (D) 4

Chapter 9
Probability

This chapter covers the following Georgia Performance Standards:

MA1D	Data Analysis and Probability	MA1D1a
		MA1D2a
		MA1D2b
		MA1D2c
		MA1D2d

9.1   Probability Terms

Probability - the branch of mathematics that calculates the chance something will or will not happen.

Independent Events - the outcome of one event does not influence the outcome of the second event.

Dependent Events - the outcome of one event does influence the outcome of the second event.

Mutually Exclusive Events - two events that have no outcomes in common. These events cannot occur at the same time.

Conditional Probability - probability that a second event will happen, given that the first event has already occurred.

Equally Likely Outcomes - all outcomes of the event have the same chance of occurring.

Expected Value - the mean of a random variable.

Population - an entire group or collection about which we wish to draw conclusions.

Census - a count of the entire population

Sample - units selected to study from the population.

Sample Space - the set of all possible outcomes.

$P(A)$ - notation used to mean the probability of outcome 'A' occurring.

$P(A^c)$ (**Complement**) - notation used to mean the probability that outcome 'A' does not occur.
$$P(A^c) = 1 - P(A)$$

9.2   Probability

Probability is the chance something will happen. Probability is most often expressed as a fraction, a decimal, a percent, or can also be written out in words.

Example 1:   Billy has 3 red marbles, 5 white marbles, and 4 blue marbles on the floor. His cat comes along and bats one marble under the chair. What is the **probability** it is a red marble?

Step 1:   The number of red marbles will be the top number of the fraction. $\longrightarrow \dfrac{3}{12}$

Step 2:   The total number of marbles is the bottom number of the fraction. $\longrightarrow$

The answer may be expressed in lowest terms. $\dfrac{3}{12} = \dfrac{1}{4}$.

Expressed as a decimal, $\frac{1}{4} = 0.25$, as a percent, $\frac{1}{4} = 25\%$, and written out in words $\frac{1}{4}$ is one out of four.

Example 2:   Determine the probability that the pointer will stop on a shaded wedge or the number 1.

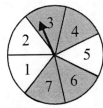

Step 1:   Count the number of possible wedges that the spinner can stop on to satisfy the above problem. There are 5 wedges that satisfy it (4 shaded wedges and one number 1). The top number of the fraction is 5.

Step 2:   Count the total number of wedges, 7. The bottom number of the fraction is 7. The answer is $\frac{5}{7}$ or **five out of seven.**

Example 3:   Refer to the spinner above. If the pointer stops on the number 7, what is the probability that it will **not** stop on 7 the next time?

Step 1:   Ignore the information that the pointer stopped on 7 the previous spin. The probability of the next spin does not depend on the outcome of the previous spin. Simply find the probability that the spinner will **not** stop on 7. Remember, if P is the probability of an event occurring, $1 - P$ is the probability of an event **not** occurring (it is the complement). In this example, the probability of the spinner landing on 7 is $\frac{1}{7}$.

Step 2:   The probability that the spinner will not stop on 7 is $1 - \frac{1}{7}$ which equals $\frac{6}{7}$. The answer is $\frac{6}{7}$ or **six out of seven.**

Find the probability of the following problems. Express the answer as a percent.

1. A computer chooses a random number between 1 and 50. What is the probability that you will guess the same number that the computer chose in 1 try?

2. There are 24 candy-coated chocolate pieces in a bag. Eight have defects in the coating that can be seen only with close inspection. What is the probability of pulling out a defective piece without looking?

3. Seven sisters have to choose which day each will wash the dishes. They put equal-sized pieces of paper in a hat, each labeled with a day of the week. What is the probability that the first sister who draws will choose a weekend day?

4. For his garden, Clay has a mixture of 12 white corn seeds, 24 yellow corn seeds, and 16 bicolor corn seeds. If he reaches for a seed without looking, what is the probability that Clay will plant a bicolor corn seed first?

5. Mom just got a new department store credit card in the mail. What is the probability that the last digit is an odd number?

6. Alex has a paper bag of cookies that holds 8 chocolate chip, 4 peanut butter, 6 butterscotch chip, and 12 ginger. Without looking, his friend John reaches in the bag for a cookie. What is the probability that the cookie is peanut butter?

7. An umpire at a little league baseball game has 14 balls in his pockets. Five of the balls are brand A, 6 are brand B, and 3 are brand C. What is the probability that the next ball he throws to the pitcher is a brand C ball?

8. What is the probability that the spinner's arrow will land on an even number?

9. The spinner in the problem above stopped on a shaded wedge on the first spin and stopped on the number 2 on the second spin. What is the probability that it will not stop on a shaded wedge or on the 2 on the third spin?

10. A company is offering 1 grand prize, 3 second place prizes, and 25 third place prizes based on a random drawing of contest entries. If your entry is one of the 500 total entries, what is the probability you will win a third place prize?

11. In the contest problem above, what is the probability that you will win the grand prize or a second place prize?

12. A box of a dozen doughnuts has 3 lemon cream-filled, 5 chocolate cream-filled, and 4 vanilla cream-filled. If the doughnuts look identical, what is the probability of picking a lemon cream-filled?

9.3 More Probability

Example 4: You have a cube with one number, 1, 2, 3, 4, 5, and 6 painted on each face of the cube. What is the probability that if you throw the cube 3 times, you will get the number 2 each time?

If you roll the cube once, you have a 1 in 6 chance of getting the number 2. If you roll the cube a second time, you again have a 1 in 6 chance of getting the number 2. If you roll the cube a third time, you again have a 1 in 6 chance of getting the number 2. The probability of rolling the number 2 three times in a row is:

$$\frac{1}{6} \times \frac{1}{6} \times \frac{1}{6} = \frac{1}{216}$$

Find the probability that each of the following events will occur.

There are 10 balls in a box, each with a different digit on it: 0, 1, 2, 3, 4, 5, 6, 7, 8, & 9. A ball is chosen at random and then put back in the box.

1. What is the probability that if you pick out a ball 3 times, you will get number 7 each time?

2. What is the probability you will pick a ball with 5, then 9, and then 3?

3. What is the probability that if you pick out a ball 4 times, you will always get an odd number?

4. A couple has 4 children ages 9, 6, 4, and 1. What is the probability that they are all girls?

There are 26 letters in the alphabet, allowing a different letter to be on each of 26 cards. The cards are shuffled. After each card is chosen at random, it is put back in the stack of cards, and the cards are shuffled again.

5. What is the probability that when you pick 3 cards, you would draw first a "y", then an "e", and then an "s"?

6. What is the probability that you would draw 4 cards and get the letter "z" each time?

7. What is the probability that you draw twice and get a letter in the word "random" both times?

8. If you flip a coin 3 times, what is the probability you will get heads every time?

9. Marie is clueless about 4 of her multiple-choice answers. The possible answers are A, B, C, D, E, or F. What is the probability that she will guess all four answers correctly?

9.4   Tree Diagrams

Drawing a tree diagram is another method of determining the probability of events occurring.

Example 5:   If you toss two six-sided numbered cubes that have 1, 2, 3, 4, 5, or 6 on each side, what is the probability you will get two cubes that add up to 9? One way to determine the probability is to make a tree diagram.

Cube 1	Cube 2	Cube 1 plus Cube 2
1	1	2
	2	3
	3	4
	4	5
	5	6
	6	7
2	1	3
	2	4
	3	5
	4	6
	5	7
	6	8
3	1	4
	2	5
	3	6
	4	7
	5	8
	6	⑨
4	1	5
	2	6
	3	7
	4	8
	5	⑨
	6	10
5	1	6
	2	7
	3	8
	4	⑨
	5	10
	6	11
6	1	7
	2	8
	3	⑨
	4	10
	5	11
	6	12

Alternative method

Write down all of the numbers on both cubes which would add up to 9.

Cube 1	Cube 2
4	5
5	4
6	3
3	6

Numerator = 4 combinations

For denominator: Multiply the number of sides on one cube times the number of sides on the other cube.

$6 \times 6 = 36$

Numerator:
Denominator: $\dfrac{4}{36} = \dfrac{1}{9}$

There are 36 possible ways the cubes could land. Out of those 36 ways, the two cubes add up to 9 only 4 times. The probability you will get two cubes that add up to 9 is $\dfrac{4}{36}$ or $\dfrac{1}{9}$.

Read each of the problems below. Then answer the questions.

1. Jake has a spinner. The spinner is divided into eight equal regions numbered 1–8. In two spins, what is the probability that the numbers added together will equal 12?

2. Charlie and Libby each spin one spinner one time. The spinner is divided into 5 equal regions numbered 1–5. What is the probability that these two spins added together would equal 7?

3. Gail spins a spinner twice. The spinner is divided into 9 equal regions numbered 1–9. In two spins, what is the probability that the difference between the two numbers will equal 4?

4. Diedra throws two 10-sided numbered polyhedrons. What is the probability that the difference between the two numbers will equal 7?

5. Cameron throws two six-sided numbered cubes. What is the probability that the difference between the two numbers will equal 3?

6. Tesla spins one spinner twice. The spinner is divided into 11 equal regions numbered 1–11. What is the probability that the two numbers added together will equal 11?

7. Samantha decides to roll two five-sided numbered prisms. What is the probability that the two numbers added together will equal 4?

8. Mary Ellen spins a spinner twice. The spinner is divided into 7 equal regions numbered 1–7. What is the probability that the product of the two numbers equals 10?

9. Conner decides to roll two six-sided numbered cubes. What is the probability that the product of the two numbers equals 4?

10. Tabitha spins one spinner twice. The spinner is divided into 9 equal regions numbered 1–9. What is the probability that the sum of the two numbers equals 10?

11. Darnell decides to roll two 15-sided numbered polyhedrons. What is the probability that the difference between the two numbers is 13?

12. Inez spins one spinner twice. The spinner is divided into 12 equal regions numbered 1–12. What is the probability that the sum of two numbers equals 10?

13. Gina spins one spinner twice. The spinner is divided into 8 equal regions numbered 1–8. What is the probability that the two numbers added together equals 9?

14. Celia rolls two six-sided numbered cubes. What is the probability that the difference between the two numbers is 2?

15. Brett spins one spinner twice. The spinner is divided into 4 equal regions numbered 1–4. What is the probability that the difference between the two numbers will be 3?

9.5   Independent and Dependent Events

In mathematics, the outcome of an event may or may not influence the outcome of a second event. If the outcome of one event does not influence the outcome of the second event, these events are **independent**. However, if one event has an influence on the second event, the events are **dependent**. When someone needs to determine the probability of two events occurring, he or she will need to use an equation. These equations will change depending on whether the events are independent or dependent in relation to each other. When finding the probability of two **independent** events, multiply the probability of each favorable outcome together. Independent events use the **multiplication rule**. The multiplication rule for independent events is $P(A \text{ and } B) = P(A) P(B)$.

Example 6:   One bag of marbles contains 1 white, 1 yellow, 2 blue, and 3 orange marbles. A second bag of marbles contains 2 white, 3 yellow, 1 blue, and 2 orange marbles. What is the probability of drawing a blue marble from each bag?

Solution: Probability of favorable outcomes

Bag 1 $P(A)$: $\dfrac{2}{7}$

Bag 2 $P(B)$: $\dfrac{1}{8}$

Probability of a blue marble from each bag $= P(A \text{ and } B)$

$$P(A \text{ and } B) = P(A) P(B) = \frac{2}{7} \times \frac{1}{8} = \frac{2}{56} = \frac{1}{28}$$

In order to find the probability of two **dependent** events, you will need to use a different set of rules. For the first event, you must divide the number of favorable outcomes by the number of possible outcomes. For the second event, you must subtract one from the number of favorable outcomes **only if** the favorable outcome is the **same**. However, you must subtract one from the number of total possible outcomes. Finally, you must multiply the probability for event one by the probability for event two.

Example 7:   One bag of marbles contains 3 red, 4 green, 7 black, and 2 yellow marbles. What is the probability of drawing a green marble, removing it from the bag, and then drawing another green marble without looking?

	Favorable Outcomes	Total Possible Outcomes
Draw 1	4	16
Draw 2	3	15
Draw 1 × Draw 2	12	240

Answer: $\dfrac{12}{240}$ or $\dfrac{1}{20}$

Example 8:   Using the same bag of marbles, what is the probability of drawing a red marble and then drawing a black marble?

	Favorable Outcomes	Total Possible Outcomes
Draw 1	3	16
Draw 2	7	15
Draw 1 × Draw 2	21	240

Answer $\dfrac{21}{240}$ or $\dfrac{7}{80}$

Find the probability of the following problems. Express the answer as a fraction.

1. Prithi has two boxes. Box one contains 3 red, 2 silver, 4 gold, and 2 blue combs. She also has a second box containing 1 black and 1 clear brush. What is the probability that Prithi selects a red comb from box one and a black brush from the second box?

2. Steve Marduke has two spinners in front of him. The first one is numbered 1–6, and the second is numbered 1–3. If Steve spins each spinner once, what is the probability that the first spinner will show an odd number and the second spinner will show a "1"?

3. Carrie McCallister flips a coin twice and gets heads both times. What is the probability that Carrie will get tails the third time she flips the coin?

4. Artie Drake turns a spinner which is evenly divided into 11 sections numbered 1–11. On the first spin, Artie's pointer lands on "8". What is the probability that the spinner lands on an even number the second time he turns the spinner?

5. Leanne Davis plays a game with a street entertainer. In this game, a ball is placed under one of three coconut halves. The vendor shifts the coconut halves so quickly that Leanne can no longer tell which coconut half contains the ball. She selects one and misses. The entertainer then shifts all three around once more and asks Leanne to pick again. What is the probability that Leanne will select the coconut half containing the ball?

6. What is the probability that Jane Robelot reaches into a bag containing 1 daffodil and 2 gladiola bulbs and pulls out a daffodil bulb, and then reaches into a second bag containing 6 tulip, 3 lily, and 2 gladiola bulbs and pulls out a lily bulb?

7. Terrell casts his line into a pond containing 7 catfish, 8 bream, 3 trout, and 6 northern pike. He immediately catches a bream. What are the chances that Terrell will catch a second bream the next time he casts his line?

8. Gloria Quintero enters a contest in which the person who draws his or her initials out of a box containing all 26 letters of the alphabet wins the grand prize. Gloria reaches in, draws a "G", keeps it, then draws another letter. What is the probability that Gloria will next draw a "Q"?

9. Vince Macaluso is pulling two socks out of a washing machine in the dark. The washing machine contains three tan, one white, and two black socks. If Vince reaches in and pulls out the socks one at a time, what is the probability that he will pull out two tan socks on his first two tries?

10. John Salome has a bag containing 2 yellow plums, 2 red plums, and 3 purple plums. What is the probability that he reaches in without looking and pulls out a yellow plum and eats it, then reaches in again without looking and pulls out a red plum to eat?

9.6   Mutually Exclusive Events

Events are said to be **mutually exclusive** if they don't occur at the same time (no common outcome). This means the probability of event 1 (A) and event 2 (B) both occurring is zero, $P(A$ and $B) = 0$. Mutually exclusive events use the **addition rule**.

Let A and B be events. Let $P(A) =$ the probability of event A occurring and $P(B) =$ the probability of event B occurring.

The addition rule for mutually exclusive events is $P(A$ or $B) = P(A) + P(B)$.

The rule for sets that are **not** mutually exclusive is $P(A$ or $B) = P(A) + P(B) - P(A \cap B)$.

Example 9:   A pair of diced is rolled. What is the probability that the sum of the dice rolled is either a 7 or a 2?

Step 1:   Find the number of outcomes for rolling a 7.
There are six outcomes for rolling a seven:
$(1, 6), (6, 1), (2, 5), (5, 2), (3, 4), (4, 3)$.

$$P(7) = \frac{\text{\# of outcomes for rolling a seven}}{\text{total number of outcomes}} = \frac{6}{36} = \frac{1}{6}$$

Step 2:   Then find the number of outcomes for rolling a 2.
There is one outcome for rolling a two:
$(1, 1)$

$$P(2) = \frac{\text{\# of outcomes for rolling a two}}{\text{total number of outcomes}} = \frac{1}{36}$$

Step 3:   Since the sum of the dice cannot be seven and two (it must be one or the other), then the events are mutually exclusive. Use the formula $P(x$ or $y) = P(x) + P(y)$ for mutually exclusive events to find the probability of either rolling a sum of 7 or a sum of 2.
Let $x =$ rolling a sum of 7 and $y =$ rolling a sum of 2

$$P(7 \text{ or } 2) = P(7) + P(2) = \frac{1}{6} + \frac{1}{36} = \frac{7}{36}$$

The probability of rolling the sum of either a 7 or a 2 is $\frac{7}{36}$.

Find the probability of each event.

1. A pair of dice is rolled. What is the probability that the sum of the dice rolled is either an 11 or a 5?

2. A pair of dice is rolled. What is the probability that the sum of the dice rolled is either a 9 or a 3?

3. A pair of dice is rolled. What is the probability that the sum of the dice rolled is either a 6 or a 12?

4. Russell has a bag of 100 marbles. 50 are red and 50 are black. What is the probability of
   (A) picking 2 red marbles if the first one is NOT replaced?
   (B) picking two black and two red marbles without replacement?

9.7   Conditional Probability

Conditional probability is defined as the probability that a second event will happen, given that the first event has already occurred. When two events are dependent, the conditional probability of A given B is

$$P\left(A \text{ given } B\right) = P(A|B) = \frac{P(A \text{ and } B)}{P(B)} = \frac{P(A \cap B)}{P(B)}.$$

Example 10:   A pair of dice is rolled. What is the probability that the sum of two dice will be greater than 7, A, if the first die rolled is a 4, B?

Step 1:   Find A and B.
$A = \text{total of two dice} > 7$
$B = 4$

Step 2:   Find the number of outcomes for the sum of the dice. There are three outcomes, since the first die must be four and the sum of the dice must be greater than 7.
$A \text{ and } B = A \cap B = (4, 4), (4, 5), (4, 6)$

Step 3:   Find $P(A \text{ and } B)$ and $P(B)$.
$$P(A \text{ and } B) = \frac{\# \text{ of outcomes for } A \text{ and } B}{\text{total } \# \text{ of outcomes}} = \frac{3}{36} = \frac{1}{12}$$
$$P(B) = \frac{\# \text{ of outcomes for } B}{\text{total } \# \text{ of outcomes}} = \frac{1}{6}$$
$$P(A|B) = \frac{P(A \text{ and } B)}{P(B)} = \frac{\frac{1}{12}}{\frac{1}{6}} = \frac{1}{2}$$

The probability of rolling the sum of two dice that is greater than 7, given that the first dice must be 4 is $\frac{1}{2}$.

Example 11:   The probability that UGA and GT both win a football game in the same weekend is 37%. The probability that just UGA wins is 68%. What is the probability that GT will win given that UGA has already won?

$$P(A|B) = \frac{P(A \text{ and } B)}{P(B)} = \frac{37\%}{68\%} = 54\%$$

Find the conditional probability.

1. Kelly took two math tests. Her teacher said 30% of the class passed both tests, but 52% passed the first test. What is the probability that Kelly, who passed the first test, also passed the second test?

2. In California, 88% of teenagers have a cell phone and 76% have a cell phone and a MP3 player. What is the probability that a teenager has a MP3 player given that he or she also has a cell phone?

3. Tonya is selling 2 kinds of cookies, chocolate chip and sugar. After going to 100 houses, she determines that 41% of people bought chocolate chip cookies and 38% bought chocolate chip and sugar cookies. What percentage of people bought sugar cookies given that they already purchased chocolate chip cookies?

4. Lenny and Rebecca bought two scratch off lottery tickets. The probability that they both win is 6%. The probability that just one of them wins is 13%. What is the probability that Lenny will win given the fact that Rebecca already won?

5. The Braves have a double-header on Saturday. The probability that they will win both games is 31%. The probability that they will win just the first game is 71%. What is the probability they will win the second game given that they have already won the first game?

6. During a storm, the probability of the power and the cable going out is 31%. The probability that just the cable will go out is 54%. What is the probability that the power will go out given that the cable is already out?

7. The Nutcracker is being performed at the Fox Theater. The probability that the show will sell out on Friday and Saturday night is 61%. The probability that it will sell out Friday night is 63%. What is the probability that the show will sell out Saturday night if it sold out Friday night?

8. During a race, the probability that a racecar driver will have two flat tires is 13%. The probability that he will have one flat tire is 21%. Given that he has already had one flat tire, what is the probability that he will have another?

9. The probability of thunder and lightning during a summer storm is 19%. The probability of only having thunder is 27%. What is the probability that lightning will occur given that there is already thunder?

10. Matthew and Lexi decided to go out to dinner. The probability that they both order a hamburger is 7%. The probability that one of them orders a hamburger is 24%. What is the probability that Matthew will order a hamburger if Lexi has already ordered one?

11. A pair of dice is rolled. What is the probability that the sum of two dice will be greater than 9 if the first die rolled is a 6?

12. A pair of dice is rolled. What is the probability that the sum of two dice will be less than 5 if the second die rolled is a 2?

9.8 Predictions Using Probabilities

Once you know the probability of an event, you can use that probability to make predictions about future events.

Example 12: Slugger Jones has a 0.475 batting average this season, meaning that he gets a hit 47.5% of his official at-bats (excluding walks, hit by pitch, etc.). In the double-header yesterday, he had 11 official at-bats. How many hits would you have expected him to collect?

Solution: Multiply the probability by the number of total opportunities. $0.475 * 11 = 5.225$ We would expect him to have gotten 5 hits in 11 at-bats.

Example 13: Farmer Jacob owns so many sheep that he has 5 assistant shepherds. He wants to check on the composition of his flock, so he selects 40 sheep at random to see what they look like. Striped sheep are the most valuable. The results of his survey are below:

Type of sheep	White	Black	Striped	Speckled
Number of sheep	15	12	7	6

If one sheep is chosen at random, what is the probability that it is striped? If Farmer Jacob has 750 sheep total, how many of his sheep will probably be striped?

Step 1: The probability of one sheep being striped is
$$\frac{\text{The number of striped sheep}}{\text{The total number of sheep}} = \frac{7}{40} = 17.5\%$$

Step 2: Multiply the probability of one sheep being striped by the total number of sheep in the flock:
$0.175 \times 750 = 131.25$
He most likely has 131 striped sheep.

Example 14: Mr. Gallop takes a poll of a randomly selected portion of the 10th grade class about their favorite baseball team. He makes a chart with the top 5 responses:

Top 5 Responses Among 10th Graders to the Question, "Who is Your Favorite Baseball Team?"

Favorite Team	Baltimore	Washington, DC	New York	Chicago	Boston
# of Responses	24	18	7	6	5

Mr. Gallop reads this chart and counts that there are 63 responses listed. He concludes that there is a $\frac{7}{63} = \frac{1}{9}$ probability that a randomly selected student would be a New York fan.
Therefore, he says, out of the 432 students in the 10th grade, $432 \times \frac{1}{9} = 48$ are likely to be New York fans. Why is his conclusion invalid?

Solution: The chart only gives the Top 5 responses about 10th graders to Mr. Gallop's question, not every response. Therefore, we do not know how many total students were polled, and we cannot calculate probabilities or make predictions.

Use the probabilities in the questions below to make predictions, if possible. If it is not possible to make predictions, explain why.

1. Mrs. Polley's oven burns about 6% of the cookies in a batch, on average. She makes 400 cookies for her daughter Lynn's wedding. How many cookies should she expect her oven to burn?

2. Christie Jo is on the Prom Planning Panel (PPP). She does a random survey of 30 female classmates to find out their plans for prom. The results are shown below.

Prom Plans	Have a Date	Plan on Attending with or without a Date	Not Attending
# of Females	5	16	9

   What is the probability that a female student selected at random is planning to attend? If there are 340 female students eligible to go to the prom, how many should Christie Jo expect to attend (based on this survey)?

3. Prisci surveyed a random sample of her classmates about their favorite type of ice cream. The results are shown below.

   Favorite Types of Ice Cream at Grosami High School

Favorite Ice Cream*	Vanilla (incl. French Vanilla)	Chocolate	Strawberry	Chocolate Chip
# of Responses	24	8	5	3

   * lists only ice cream flavors that received 3 or more responses

   According to the survey results, what is the probability that a randomly selected student would prefer chocolate? If there are 80 students total in Prisci's class, how many of them would she predict prefer chocolate?

4. Van uses the SureFire cell phone company. They claim to have a dropped call rate of 5% in his area. If Van plans on making 60 phone calls this month, how many dropped calls should he expect?

5. Peggy takes a lot of photographs, and she prefers the quality of old-style film cameras. She buys rolls of film that take 60 pictures each. The table below shows the number of photos from each roll that Peggy rejects from putting in a photo album.

   Peggy's Rejected Photos (rolls of 60)

Film Roll	1	2	3	4	5	6	7
Rejected Photos	3	8	2	0	4	5	5

   In a year, Peggy uses 40 rolls, or 2400 photos. Based on these 7 rolls, how many photos would you expect her to reject? Round to the nearest whole number.

9.9   Expected Value

The expected value of a variable is the same as the average or mean of the variable. If we have a variable, X, the expected value is $E(X) = \sum_{i=1}^{n} X_i P_i$ where P_i is the probability of the value of X_i.

Example 15:   Suppose that the following game is played. A man rolls a fair die. If he rolls 1, 3, or 5, he loses \$3, if he rolls 4 or 6, he loses \$2, and if he rolls 2, he wins \$12. What gain (loss) should he expect on average? (What is his expected value?)

Step 1:   We can define a random variable X to be the amount of money associated with each outcome.

Therefore, the possible values of X are $-\$3$ (if 1, 3, or 5 is rolled), $-\$2$ (if 4 or 6 is rolled), and \$12 (if 2 is rolled).

Note: \$3 and \$2 are negative because he loses that money, not gains.

Step 2:   Find the probabilities of each of the outcomes, $P(-\$3)$, $P(-\$2)$, and $P(\$12)$. To do this make a table.

Possible values of X	Probability $P(X)$
$-\$3$	$P(\text{rolling a 1, 3, or 5}) = \frac{3}{6}$
$-\$2$	$P(\text{rolling a 4 or 6}) = \frac{2}{6}$
$\$12$	$P(\text{rolling a 2}) = \frac{1}{6}$

Step 3:   Now, we multiply the probability for each outcome by the amount of money associated with that specific outcome.

Expected Value of Rolling a 1, 3, or, 5: $-\$3 \times \frac{3}{6} = -\$1.5$

Expected Value of Rolling a 4 or 6: $-\$2 \times \frac{2}{6} = -\$0.67$

Expected Value of Rolling a 2: $\$12 \times \frac{1}{6} = \$2$

Step 4:   To find the expected value of the entire game, add up the expected value for each outcome.

$$E(X) = (-\$3)\left(\tfrac{3}{6}\right) + (-\$2)\left(\tfrac{2}{6}\right) + (\$12)\left(\tfrac{1}{6}\right)$$

$$E(X) = -\$1.5 - \$\tfrac{2}{3} + \$2 = -\$\tfrac{1}{6}$$

The expected value is $-\$\frac{1}{6}$.

This means that the man playing the game is expected to lose an average of 16.7 cents $\left(\frac{1}{6} \text{ of } \$1.00\right)$ each game he plays.

Find the expected value.

1. Ken tells you to throw a die 100 times and he will give you whatever you roll in dollars (if you roll a 1, you get $1 etc.). Before you play, you want to estimate how much money Ken will end up giving you.

2. Mark and Rodney are playing a card game. Every time Mark draws a Jack or Queen, Rodney gives him $4. When Mark draws a King or Ace, Rodney gives him $5. In turn, Mark gives Rodney $X for every other card drawn. In order for the game to be fair (the expected value is zero), how much should Mark pay Rodney for the cards he draws? (Hint: Find the value of X.)

3. On the first day of class, Kara's teacher gave the class the following chart to represent the test number and the probable percent of people that will make an A on each test.

Test	1	2	3	4	5	6	7	8	9	10
P(A)	0.36	0.32	0.28	0.25	0.21	0.19	0.16	0.14	0.09	0.03

How many people out of a class of 40 will make an A? Round to the nearest whole number.

4. Suppose that the following game is played. A man rolls a fair die. If he rolls 2 or 5, he loses $5, if he rolls 1 or 3, he loses $7, and if he rolls 4 or 6, he wins $15. What gain (loss) should he expect on average?

5. Suppose that the following game is played. A woman rolls a fair die. If she rolls 1, she loses $15, if she rolls 2 or 3, she loses $10, and if she rolls 4, 5, or 6, she wins $20. What gain (loss) should she expect on average?

6. Suppose that the following game is played. A man rolls a fair die. If he rolls 1 or 5, he loses $3, if he rolls 2, 3, or 6, he loses $5, and if he rolls 4, he wins $11. What gain (loss) should he expect on average?

7. Suppose that the following game is played. A woman rolls a fair die. If she rolls 1 or 6, she loses $1, if she rolls 2 or 5, she loses $2, and if she rolls 3 or 4, she wins $9. What gain (loss) should she expect on average?

8. Suppose that the following game is played. A man rolls a fair die. If he rolls 1, he loses $6, if he rolls 2, 3, or 4, he loses $3, if he rolls 5, he loses $5, and if he rolls 6, he wins $18. What gain (loss) should he expect on average?

Chapter 9 Review

1. There are 50 students in the school orchestra in the following sections:

string section	woodwind	percussion	brass
25	15	5	5

   One student will be chosen at random to present the orchestra director with an award. What is the probability the student will be from the woodwind section?

2. Fluffy's cat treat box contains 6 chicken-flavored treats, 5 beef-flavored treats, and 7 fish-flavored treats. If Fluffy's owner reaches in the box without looking, and chooses one treat, what is the probability that Fluffy will get a chicken-flavored treat?

3. The spinner in figure A stopped on the number 5 on the first spin. What is the probability that it will not stop on 5 on the second spin?

   Fig. A   Fig. B

4. Sherri turns the spinner in figure B above 3 times. What is the probability that the pointer always lands on a shaded number?

5. Three cakes are sliced into 20 pieces each. Each cake contains 1 gold ring. What is the probability that one person who eats one piece of cake from each of the 3 cakes will find 3 gold rings?

6. Brianna tosses a coin 4 times. What is the probability she gets all tails?

7. Tempest has a bag with 4 red marbles, 3 blue marbles, and 2 yellow marbles. Does adding 4 purple marbles increase or decrease her chances that the first marble she draws at random will be red?

8. Simone has lived in Silver Spring for 250 days, and in that time her power has been out 3 days.

   (A) If the power outages happen at random, what is the probability that the power will be out tomorrow?

   (B) If the probability remains the same, how many days will she be without power for the next 10 years (3, 652 days)? Round to the nearest whole number.

9. There are 20 balls in a box. 12 are green, 8 are black. If two balls are chosen without replacement, what is the probability that both are green balls?

10. Suppose that the following game is played. A man rolls a fair die. If he rolls 2, 4, or 6, he loses $2, if he rolls 1 or 3, he loses $5, and if he rolls 5, he wins $9. What gain (loss) should he expect on average? (What is his expected value?)

11. During football season, the probability of students missing school is 16%. The probability that they will miss school and go to the football game is 9%. What is the probability a student will go to the game if they miss school on Friday?

12. At Texas State University, the probability that a student is taking Spanish is 72%. The probability that a student is taking Spanish and Self Defense is 28%. What is the probability that a student is taking Self Defense given that they are already registered to take Spanish?

13. Branden has a box of 50 cookies, 4 of which are peanut butter. On the way home he randomly picks two cookies to eat as his snack. What is the probability he chose

(A) 2 peanut butter cookies (no replacement)?
(B) 2 peanut butter cookies if he puts the first one back?
(C) 1 peanut butter cookie and 1 other cookie (no replacement)?

14. At the county fair, there are 5 unlabeled cakes. 3 are on table A. 2 of those 3 were made by Ann and the third was made by Betty. Two cakes (one of Ann's and one of Betty's) are on table B. The judge at the fair moves one cake from table A to table B. Jack then randomly chooses a cake from table B. What is the probability that he chose a cake made by Betty?

Read the following, and answer questions 15–19.

There are 9 slips of paper in a hat, each with a number from 1 to 9. The numbers correspond to a group of students who must answer a question when the number for their group is drawn. Each time a number is drawn, the number is put back in the hat.

15. What is the probability that the number 6 will be drawn twice in a row?

16. What is the probability that the first 5 numbers drawn will be odd numbers?

17. What is the probability that the second, third, and fourth numbers drawn will be even numbers?

18. What is the probability that the first five times a number is drawn it will be the number 5?

19. What is the probability that the first five numbers drawn will be 1, 2, 3, 4, 5 in that order?

Solve the following word problems. For questions 20–22, write whether the problem is "dependent" or "independent".

20. Felix Perez reaches into a 10-piece puzzle and pulls out one piece at random. This piece has two places where it could connect to other pieces. What is the probability that he will select another piece which fits the first one if he selects the next piece at random?

21. Barbara Stein is desperate for a piece of chocolate candy. She reaches into a bag which contains 8 peppermint, 5 butterscotch, 7 toffee, 3 mint, and 6 chocolate pieces and pulls out a toffee piece. Disappointed, she throws it back into the bag and then reaches back in and pulls out one piece of candy. What is the probability that Barbara pulls out a chocolate piece on the second try?

22. Christen Solis goes to a pet shop and immediately decides to purchase a guppy she saw swimming in an aquarium. She reaches into the tank containing 5 goldfish, 6 guppies, 4 miniature catfish, and 3 minnows and accidently pulls up a goldfish. Breathing a sigh, Christen places the goldfish back in the water. The fish are swimming so fast, it is impossible to tell what fish Christen would catch. What is the probability that Christen will catch a guppy on her second try?

Chapter 9 Test

1. There are 10 boys and 12 girls in a class. If one student is selected at random from the class, what is the probability it is a girl?

   (A) $\dfrac{1}{2}$

   (B) $\dfrac{1}{22}$

   (C) $\dfrac{6}{11}$

   (D) $\dfrac{6}{5}$

2. David just got a new credit card in the mail. What is the probability the second digit of the credit card number is a 3?

   (A) $\dfrac{1}{3}$

   (B) $\dfrac{1}{10}$

   (C) $\dfrac{2}{3}$

   (D) $\dfrac{1}{5}$

3. Brenda has 18 fish in an aquarium. The fish are the following colors: 5 orange, 7 blue, 2 black, and 4 green. Brenda also has a trouble-making cat that has grabbed a fish. What is the probability the cat grabbed a green fish if all the fish are equally capable of avoiding the cat?

   (A) $\dfrac{2}{9}$

   (B) $\dfrac{1}{18}$

   (C) $\dfrac{2}{7}$

   (D) $\dfrac{1}{4}$

4. In problem number 3, what is the probability the cat will **not** grab an orange fish?

   (A) $\dfrac{1}{3}$

   (B) $\dfrac{13}{18}$

   (C) $\dfrac{3}{4}$

   (D) $\dfrac{13}{5}$

5. You have a cube with each face numbered 1, 2, 3, 4, 5, or 6. What is the probability if you roll the cube 4 times, you will get the number 5 each time?

   (A) $\dfrac{4}{1296}$

   (B) $\dfrac{4}{5}$

   (C) $\dfrac{1}{256}$

   (D) $\dfrac{1}{1296}$

6. In a game using two numbered cubes, what is the probability of **not** rolling the same number on both cubes three times in a row?

   (A) $\dfrac{125}{256}$

   (B) $\dfrac{125}{216}$

   (C) $\dfrac{27}{64}$

   (D) $\dfrac{16}{27}$

7. Carrie bought a large basket of 60 apples. When she got them home, she found 4 of the apples were rotten. If she goes back and buys 200 more apples, about how many rotten apples should she expect.

(A) 8
(B) 13
(C) 20
(D) 24

8. Katie spun a spinner 15 times and recorded her results in a table below. The spinner was divided into 6 sections numbered 1–6. The results of the spins are shown below.

$$
\begin{array}{ccccc}
1 & 3 & 6 & 5 & 1 \\
6 & 2 & 4 & 3 & 4 \\
2 & 5 & 1 & 4 & 5 \\
\end{array}
$$

Based on the results, how many times would 4 be expected to appear in 45 spins?

(A) 9
(B) 12
(C) 15
(D) 21

9. Ben has a bag of 40 red triangles and 40 blue triangles. Without replacement, about what is the probability of Ben choosing 2 red triangles?

(A) 0.20
(B) 0.25
(C) 0.30
(D) 0.35

10. Kyle has 4 different kinds of Halloween candy in a bag. He has 5 chocolate rolls, 6 lollipops, 4 chocolate bars, and 7 peanut butter candies. What kind of candy must Kyle have picked out of the bag if the probability of picking that kind is $\frac{2}{11}$?

(A) chocolate roll
(B) lollipop
(C) chocolate bar
(D) peanut butter

11. There are 7 sections of equal size on a spinner. One is labeled purple. Two are labeled green. Three are labeled orange. One section is unlabeled. What color must the unlabeled section be if the probability of spinning that color is $\frac{4}{7}$?

(A) purple
(B) green
(C) orange
(D) brown

12. The table below shows the actual sum of the rolling of two cubes numbered 1 through 6. The two cubes were rolled 100 times.

Sum	Frequency
2	5
3	5
4	9
5	10
6	15
7	14
8	15
9	11
10	9
11	4
12	3

Using the information in the table, predict how many times a score of "7" would occur in 150 tries.

(A) 21
(B) 28
(C) 35
(D) 43

13. In Florida, 94% of people have a TV. 71% have a TV and a VCR. What percent of people have a VCR if they already had a TV?

(A) 66%
(B) 70%
(C) 83%
(D) 76%

14. If there are ten blocks in a bucket, 7 are red and 3 are yellow, what is the probability of choosing 2 red blocks (without replacement)?

   (A) $\dfrac{8}{15}$

   (B) $\dfrac{7}{30}$

   (C) $\dfrac{6}{15}$

   (D) $\dfrac{7}{15}$

15. If there are 100 blocks in a bucket, 40 are black, 35 are white, 20 are silver, and 5 are clear, what is the probability of choosing 6 clear blocks?

   (A) 0

   (B) $\dfrac{1}{2}$

   (C) $\dfrac{1}{20}$

   (D) $\dfrac{1}{10}$

16. Karen has two tests this week. The chance that she'll pass both is 62%. The chance she'll pass one is 74%. What's the chance she'll pass one if she already passes the other?

   (A) 80%

   (B) 87%

   (C) 89%

   (D) 84%

17. In high school, 23% of boys play football. 16% play football and basketball. What percent of boys play basketball if they are already playing football?

   (A) 50%
   (B) 70%
   (C) 80%
   (D) 30%

18. Sarah has a bag of 100 marbles. 40 are blue, 40 are green, and 20 are purple. What is the probability of Sarah picking a purple marble on the first try?

   (A) $\dfrac{1}{5}$

   (B) $\dfrac{1}{4}$

   (C) $\dfrac{1}{100}$

   (D) $\dfrac{8}{25}$

Chapter 10
Statistics

This chapter covers the following Georgia Performance Standards:

MA1D	Data Analysis and Probability	MA1D3a
		MA1D3b
		MA1D3c
		MA1D4

10.1   Mean

In statistics, the arithmetic mean is the same as the average. To find the arithmetic mean of a list of numbers, first add together all of the numbers in the list, and then divide by the number of items in the list.

Example 1:   Find the mean of 38, 72, 110, 548.

Step 1:   First add: $38 + 72 + 110 + 548 = 768$

Step 2:   There are 4 numbers in the list so divide the total by 4. $768 \div 4 = 192$
The mean is 192.

Practice finding the mean (average). Round to the nearest tenth if necessary.

1. Dinners served:
   489 561 522 450

2. Prices paid for shirts:
   $4.89 $9.97 $5.90 $8.64

3. Piglets born:
   23 19 15 21 22

4. Student absences:
   6 5 13 8 9 12 7

5. Paychecks:
   $89.56 $99.99 $56.54

6. Choir attendance:
   56 45 97 66 70

7. Long distance calls:
   33 14 24 21 19

8. Train boxcars:
   56 55 48 61 51

9. Cookies eaten:
   5 6 8 9 2 4 3

Find the mean (average) of the following word problems.

10. Val's science grades are 95, 87, 65, 94, 78, and 97. What is her average?

11. Ann runs a business from her home. The number of orders for the last 7 business days are 17, 24, 13, 8, 11, 15, and 9. What is the average number of orders per day?

12. Melissa tracks the number of phone calls she has per day: 8, 2, 5, 4, 7, 3, 6, 1. What is the average number of calls she receives?

13. The Cheese Shop tracks the number of lunches they serve this week: 42, 55, 36, 41, 38, 33, and 46. What is the average number of lunches served?

14. Leah drives 364 miles in 7 hours. What is her average miles per hour?

15. Tim saves $680 in 8 months. How much does his savings average each month?

16. Ken makes 117 passes in 13 games. How many passes does he average per game?

142

10.2   Finding Data Missing From the Mean

Example 2:   Mara knew she had an 88% average in her biology class, but she lost one of her papers. The three papers she could find had scores of 98%, 84%, and 90%. What was the score on her fourth paper?

Step 1:   Calculate the total score on four papers with an 88% average. $0.88 \times 4 = 3.52$

Step 2:   Add together the scores from the three papers you have. $0.98 + 0.84 + 0.9 = 2.72$

Step 3:   Subtract the scores you know from the total score. $3.52 - 2.72 = 0.80$. She had 80% on her fourth paper.

Find the data missing from the following problems.

1. Gabriel earns 87% on his first geography test. He wants to keep a 92% average. What does he need to get on his next test to bring his average up?

2. Rian earned $68.00 on Monday. How much money must she earn on Tuesday to have an average of $80 earned for the two days?

3. Haley, Chuck, Dana, and Chris enter a contest to see who could bake the most chocolate chip cookies in an hour. They bake an average of 75 cookies. Haley bakes 55, Chuck bakes 70, and Dana bakes 90. How many does Chris bake?

10.3   Median

In a list of numbers ordered from lowest to highest, the **median** is the middle number. To find the **median**, first arrange the numbers in numerical order. If there is an odd number of items in the list, the **median** is the middle number. If there is an even number of items in the list, the **median** is the **average of the two middle numbers.**

Example 3:   Find the median of 42, 35, 45, 37, and 41.

Step 1:   Arrange the numbers in numerical order: 35 37 ⟦41⟧ 42 45

Step 2:   Find the middle number. The median is 41.

Example 4:   Find the median of 14, 53, 42, 6, 14, and 46.

Step 1:   Arrange the numbers in numerical order: 6 14 ⟦14 42⟧ 46 53.

Step 2:   Find the average of the two middle numbers. $(14 + 42) \div 2 = 28$. The median is 28.

Circle the median in each list of numbers.

1. 35, 55, 40, 30, and 45

2. 7, 2, 3, 6, 5, 1, and 8

3. 65, 42, 60, 46, and 90

4. 15, 16, 19, 25, 20

5. 75, 98, 87, 65, 82, 88, 100

6. 33, 42, 50, 22, and 19

7. 10, 8, 21, 14, 9, and 12

8. 43, 36, 20, and 40

9. 5, 24, 9, 18, 12, and 3

10. 48, 13, 54, 82, 90, and 7

11. 23, 21, 36, and 27

12. 9, 4, 3, 1, 6, 2, 10, and 12

10.4   Quartiles and Extremes

In statistics, large sets of data are separated into four equal parts. These parts are called **quartiles**. The **median** separates the data into two halves. Then, the median of the upper half is the **upper quartile**, and the median of the lower half is the **lower quartile**. The distance between the upper quartile and the lower quartile is the **interquartile range**. The interquartile range is sometimes used in the place of range, especially when there are outliers in the data set. Interquartile range is also another type of variability of the data.

The **extremes** are the highest and lowest values in a set of data. The lowest value is called the **lower extreme**, and the highest value is called the **upper extreme**.

Example 5:   The following set of data shows the high temperatures (in degrees Fahrenheit) in cities across the United States on a particular autumn day. Find the median, the upper quartile, the lower quartile, the upper extreme, and the lower extreme of the data.

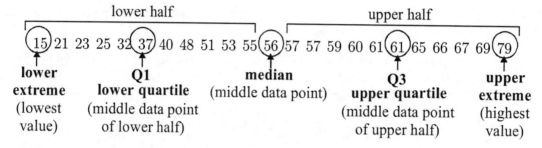

Example 6:   The following set of data shows the fastest race car qualifying speeds in miles per hour. Find the median, the upper quartile, the lower quartile, the upper extreme, and the lower extreme of the data.

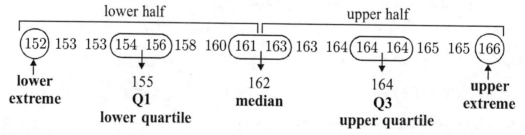

Note: When you have an even number of data points, the median is the average of the two middle points. The lower middle number is then included in the lower half of the data, and the upper middle number is included in the upper half.

Find the median, the upper quartile, the lower quartile, the upper extreme, the lower extreme, and interquartile range of each set of data given below.

1. 0 0 1 1 1 2 2 3 3 4 5

2. 15 16 18 20 22 22 23

3. 62 75 77 80 81 85 87 91 94

4. 74 74 76 76 77 78

5. 3 3 3 5 5 6 6 7 7 7 8 8

6. 190 191 192 192 194 195 196

7. 6 7 9 9 10 10 11 13 15

8. 21 22 24 25 27 28 32 35

10.5   Comparing Statistics

The best way to compare statistics is to use summary statistics. To compare, you could use the mean, median, mode, range, interquartile range, or quartiles.

Example 7:   Compare Zack and Cody's test scores using the mean, median, mode, range, interquartile range, and quartiles.

Zack	81	72	91	88	71	73	82	100	81	86
Cody	86	80	86	84	72	80	90	85	89	92

Compare the **mean** of Zack and Cody's test scores. Whose average is higher?

Zack: $\dfrac{81 + 72 + 91 + 88 + 71 + 73 + 82 + 100 + 81 + 86}{10} = \dfrac{825}{10} = 82.5$

Cody: $\dfrac{86 + 80 + 86 + 84 + 72 + 80 + 90 + 85 + 89 + 92}{10} = \dfrac{844}{10} = 84.4$

Cody's test average is higher.

Note: The mean (average) is the only value that is sensitive to extreme values. The process of comparing averages is used for samples that are very large. In cases where the sample is very large, we call the estimate of that data a **parameter**. In cases where we take the time to measure the entire population, the parameter becomes known as the **population parameter**.

Now compare their test scores using the **median**.
Zack's median = 81.5   Cody's median = 85.5
Cody's average is still higher!

Now compare their test scores using the **mode**.
Zack's mode = 81, because it appears twice in his test scores
Cody does not have a MODE because he has two sets of numbers that both appear twice (80, 86)

Now compare their test scores using the **range**.
Zack's range = 100 − 71 = 29   Cody's range = 92 − 71 = 20
Cody's range is smaller meaning his grades are a little more localized around one point, or grade.

Now compare their test scores using the **interquartile range**. (Remember: The interquartile range (IQR) is the measure of variance between the 3rd and 1st quartiles.)

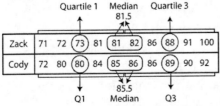

Zack's IQR = Q3 − Q1 = 88 − 73 = 15. There is a 15 point spread (variability) in grades.
Cody's IQR = Q3 − Q1 = 89 − 80 = 9. There is a 9 point spread (variability) in grades.

It is important to know that the mean, median, and mode are measures of the **center**. The range and interquartile range are measures of the **spread (variability)**.

Statistics can also be used to compare a small group of data to a large group of data.

Example 8: Stephan is a student at North Cobb High School. Look at the following histograms. The first histogram displays the amount of money earned each week by each student (rounded to the nearest ten) in Stephan's English class. The second histogram displays the amount of money earned each week by each of the students (rounded to the nearest ten) in the ninth-grade class at his school.

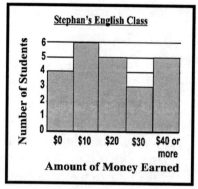 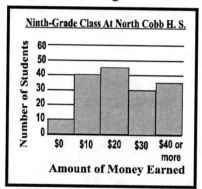

a. Compare the medians of the data in the histograms.
b. Compare the lower quartiles of the data in the histograms.
c. Compare the modes of the data in the histograms.

Solution:

a. To find the median of Stephan's English class, first find the total number of students in his class. By adding up the number of students in each bar $(4 + 6 + 5 + 3 + 5)$, we find there are 23 students in Stephan's English class. The median is the middle number in the data set that is written from least to greatest. The middle number is Stephan's English class is 12, which is found within the third bar ($20).

The total number of students in ninth-grade at North Cobb High School is found by adding up the number of students in each bar $(10 + 40 + 45 + 40 + 35)$. There are 170 students in ninth-grade at North Cobb High School. The median is between the 85th and 86th student, which is found within the third bar ($20).

The median of the amount of money earned by each of the students in the ninth-grade class at North Cobb High School is the same as the median of the amount of money earned by each of the students in Stephan's English class.

b. The lower quartile is the median of the first half of data.
The first half of data in Stephan's English class includes the first 12 students, so the median of this is found within the second bar ($10).
The first half of data in ninth-grade at North Cobb High School includes the first 85 students, so the median of this is found with the second bar ($10).
The lower quartile of the amount of money earned by each of the students in the ninth-grade class at North Cobb High School is the same as the lower quartile of the amount of money earned by each of the students in Stephan's English class.

c. The mode is easy to find. The mode is the value that is list the most.
   In Stephen's English class, the mode is $10.
   In the ninth-grade class at North Cobb High School, the mode is $20.
   The mode of the amount of money earned by each of the students in the ninth-grade class at North Cobb High School is greater than the mode of the amount of money earned by each of the students in Stephen's English class.

Use the chart below for problems 1–5.

Sample A	60	71	73	69	80	82
Sample B	71	74	73	79	81	80

1. Find the means.

2. Find the modes.

3. Find the medians.

4. Find the ranges.

5. Find the interquartile range.

Use the bar graph below for problems 6–10.

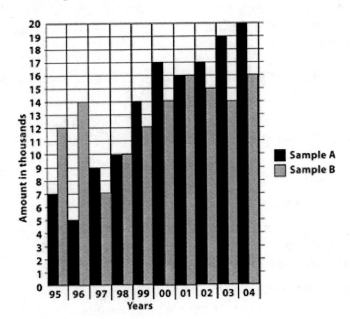

6. Find the means.

7. Find the modes.

8. Find the medians.

9. Find the ranges.

10. Find the interquartile range.

10.6   Simple Random Sampling

An important sampling technique is called **simple random sampling**. Simple random sampling is often used because it provides an excellent **representative sample** of the overall population. A representative sample is a sample that reflects (or represents) the characteristics of the whole population. There are two elements that make a sample a "simple random sample."

The first is that the sample is **unbiased**. In an unbiased sample, each member of the population is equally likely to be chosen, so the sample is representative of the population.

The second element of simple random sampling is that the observations are **independent**. In mathematics, the outcome of an event may or may not influence the outcome of a second event. If the outcome of one event does not influence the outcome of the second event these events are independent. However, if one event has an influence on the second event, the events are **dependent**.

Example 9:   Jeremy wanted to take a simple random sample of the 9th grade's movie preferences. To do so, he gave everyone in the 9th grade a ticket with a number on it, and randomly chose 15 numbers out of a hat. Then Jeremy had each of them fill out a slip of paper with their favorite movie, and he had each of them get one friend to fill out a slip with their favorite movie. Was this a simple random sample?

Solution: No, it is not a simple random sample, because the observations are not independent. The choice of the 15 friends is dependent on the 15 students who were drawn out of the hat. Jeremy should simply draw all 30 names out of the hat.

Determine whether the following samples are simple random samples, and if not, whether they are biased or dependent.

1. Henry Higgins K–8 School conducts a survey to determine children's attitudes about their parents. They get out a list of students in alphabetical order and choose every 7th name on the list to survey.

2. Million Milk tests how long it takes their milk to spoil. They use a random number generator to select 100 one-gallon bottles from among all of their products to test.

3. Reverend Nance conducts a survey on American attitudes about religion. He assigns everyone who comes to his church every Sunday a number. The reverend uses a random number generator to select 150 of them to take the survey.

4. A social scientist conducts a study on the length of time that Baltimore County newlyweds of last year dated before they married. She has the phone numbers of everyone who obtained a marriage license from Baltimore County courthouse last year, and she programs her computer to select at random 100 couples to call and survey.

Answer the following question in your own words.

5. The Cuckoo Cola Company wants to gain a sense of their workers' satisfaction, so they decide to survey some of their 17, 000 U.S. employees. Based on simple random sample principles, explain at least two of the key elements that should be part of the administration's design.

10.7   Sample Size

A final important consideration for sampling is sample size. The general rule is, the bigger the sample, the more accurate the results (as long as the sample is a representative sample).

A good rule of thumb is that for populations of 50 or smaller, you want a simple random sample size of at least $\frac{3}{4}$ or 75% of the population.

For populations larger than 50 but smaller than 100, you want a simple random sample size of at least $\frac{1}{2}$ or 50%.

For populations larger than 100, a simple random sample size of 50 or 60 is the least you want.

These rules of thumb are not exact but will help you if a question asks if the "sample size is too small." Some questions will ask you to design an experiment, and you should always include that "sample size should be large enough" and sometimes specify a sample size that is large enough.

Example 10:	Eustace came to his teacher, saying, "Mr. Wehrenberg, I just did a simple random sample and found that 50% of our class hates our school and wants the principal fired!" Mr. Wehrenberg asked Eustace how many people he had surveyed. Eustace responded, "Two."
	There are 32 people in the class. How many should Eustace survey to have a sufficient sample size, according to the rule of thumb?
	There are 700 people in the school. How many of those would Eustace have to survey to have a sufficient sample size?
Step 1:	The class has 32 students, which is less than 50, so according to the rule of thumb, he should want 75% which would be 24 students.
Step 2:	For the whole school, he should survey at least 50 or 60 students, although the more he could survey, the more accurate his results would be.

Answer the following questions.

1. True or false. Increasing the size of a simple random sample always increases its accuracy.

2. True or false. For a population of 78, a sample size of 20 would be as much as you would want for accuracy

3. True or false. For a population of 43, you should be content with a sample size of 35.

4. True or false. For a population of 50,000, a sample size of 100 should be fairly accurate.

10.8   Representative Sampling

Another important sampling technique is called **representative sampling**. Often, taking a simple random sample is not possible, either because there is no way to obtain a totally unbiased sample or a totally independent sample. The goal is still to obtain a sample as representative of the population as possible. A key element of representative sampling is **variety**.

Example 11:   Emma works for a company that produces stuffed animals at 5 factories across the Midwest. She wants to test the quality of the stuffed animals just as they come off the factory line (to eliminate the possibility of damage during shipping). Which of the following experiment designs would give the most representative sample?

(A) Go to the nearest factory and choose 100 stuffed animals at random to test.

(B) Go to each of the 5 factories and take the first 20 stuffed animals that come off the factory line.

(C) Go to each of the 5 factories and choose 20 stuffed animals at random to test.

Solution: Option (A) is a random sample that gives an excellent representation of the population of stuffed animals at that particular factory. However, Emma wants a sample that represents the population of stuffed animals at all 5 factories.

Option (C), on the other hand, is a sample that is drawn from all 5 factories. It has greater variety, so it will be more representative.

Option (B) also draws from all 5 factories, but by choosing consecutive stuffed animals, it is not an independent sample. For instance, the machines and workers may produce better stuffed animals early in the day or the week.

Option (C) is the most representative sample.

Answer the following question about representative sampling and explain your answer.

1. ReallyGoodJobs.com conducts a survey to determine the attitudes of college seniors about entering the workforce. Which of these sampling techniques would give them the most representative samples?

(A) Send surveyors to 15 randomly selected college campuses across the country and ask every person who attends the school's women's soccer game to fill out a survey.

(B) Send surveyors to interview all 600 seniors at NW Maryland State University.

(C) Choose 15 randomly selected college campuses across the country and use a computer to randomly select 40 seniors from each school out of the school's student directory.

(D) Put the survey on their website and wait until 600 college seniors respond to it.

10.9 Mean Absolute Deviation

The **mean absolute deviation** is used as a way to determine the variability of data. **Variability** (or variance) is the measure of the differences in things (scores, results, other variables, etc.) with the mean as a reference. To find the mean absolute deviation, you must compute the average of the sum of the absolute values of the deviations of the data. The deviation of element in a data set is the difference between the element and the mean of the data set. The formula for average absolute deviation for a set $\{x_1, x_2, ..., x_n\}$ is

$$\frac{1}{n} \sum_{i=1}^{n} |x_i - \overline{x}|,$$

where x_i are the elements in the set $\{x_1, x_2, ..., x_n\}$ and $\overline{x}$ is the mean of the data set.

Example 12: Find the mean absolute deviation of the data set $\{10, 10, 40, 60, 80\}$.

Step 1: First find the mean $\overline{x}$ of the data set.
$$\overline{x} = \frac{10 + 10 + 40 + 60 + 80}{5} = \frac{200}{5} = 40$$

Step 2: Now, plug the values into the equation $\frac{1}{n} \sum_{i=1}^{n} |x_i - \overline{x}|$.

$$\frac{\sum_{i=1}^{n} |x_i - \overline{x}|}{n} = \frac{|10 - 40| + |10 - 40| + |40 - 40| + |60 - 40| + |80 - 40|}{5}$$

$$= \frac{|-30| + |-30| + |0| + |20| + |40|}{5} = \frac{30 + 30 + 0 + 20 + 40}{5} = \frac{120}{5} = 24$$

The mean absolute deviation of the data set is 24.
This means that the average distance between the elements and the mean of the data set is 24. (Some elements are closer to the mean and some are further away, but the average distance is 24.)

Given a data, find the mean absolute deviation.

1. $\{1, 2, 3, 4, 5\}$

2. $\{2, 6, 8, 8\}$

3. $\{0, 1, 1, 1, 4, 5\}$

4. $\{7, 14, 15\}$

5. $\{0, 2, 4, 4, 5\}$

6. $\{0, 10, 11\}$

7. $\{2, 2, 3, 5, 8\}$

8. $\{19, 20, 21\}$

9. $\{2, 8, 13, 17\}$

10. $\{2, 3, 5, 10, 15\}$

Chapter 10 Review

Find the mean, median, mode, and range for each of the following sets of data. Fill in the table below.

❶ Miles Run by Track Team Members

Jeff	24
Eric	20
Craig	19
Simon	20
Elijah	25
Rich	19
Marcus	20

❷ 1992 SUMMER OLYMPIC GAMES
Gold Medals Won

Unified Team	45	Hungary	11
United States	37	South Korea	12
Germany	33	France	8
China	16	Australia	7
Cuba	14	Japan	3
Spain	13		

❸ Hardware Store Payroll June Week 2

Erica	$280
Dane	$206
Sam	$240
Nancy	$404
Elsie	$210
Gail	$305
David	$280

Data Set Number	Mean	Median	Mode	Range
❶				
❷				
❸				

4. Jenica bowls three games and scores an average of 116 points per game. She scores 105 on her first game and 128 on her second game. What does she score on her third game?

5. Concession stand sales for each game in season are $320, $540, $230, $450, $280, and $580. What is the mean sales per game?

6. Cendrick D'Amitrano works Friday and Saturday delivering pizza. He delivers 8 pizzas on Friday. How many pizzas must he deliver on Saturday to average 11 pizzas per day?

7. Long cooks three Vietnamese dinners that weigh a total of 40 ounces. What is the average weight for each dinner?

8. The Swamp Foxes scored an average of 7 points per soccer game. They scored 9 points in the first game, 4 points in the second game, and 5 points in the third game. What was their score for their fourth game?

9. Shondra is 66 inches tall, and DeWayne is 72 inches tall. How tall is Michael if the average height of these three students is 67 inches?

Nine cooks are asked, "If you use a thermometer, what is the actual temperature inside your oven when it is set at 350°F?" The responses are in the chart below.

Temperature (°F)	104	347	348	349	350	351	352
Number of Cooks	1	1	1	2	1	2	1

10. Find the mean of the data above.

11. Find the median of the data above.

12. Captain Kendrick of the Starship Supernova wants to know how many of his crew of 700 are afraid of encountering intelligent aliens. Which of the following would give him a simple random sample?

   (A) Ask all of his senior officers and friends.
   (B) Tell the ship's computer to assign a number between 1–700 to every crewperson and use the computer to generate random numbers to select 70 of the crew for the survey.
   (C) Get out the alphabetical crew listing and interview every 10th crew member.
   (D) Choose 10 crew members randomly from each of the 7 halls in which his crew lives.

13. True or false? Increasing the sample size of a biased sample corrects for the bias.

14. A tree farmer has thousands of trees and cannot check the health of every one regularly. He wants to design a test to check the health of some of his trees in a way that they represent the whole population of trees. What are three considerations that should go into his test design?

Use the following to answer questions 15–20.

Over the past 2 years, Coach Strive has kept a record of how many points his basketball team, the Bearcats, has scored in 20 games:

   29 32 35 36 38 39 40 40 41 42 43 44 44 45 47 49 50 52 53 62

15. What is the median?

16. What is the upper quartile?

17. What is the lower quartile?

18. What is the upper extreme?

19. What is the lower extreme?

20. What is the interquartile range?

Use the figure to help answer questions 21–27.

Jessica	92	85	97	87	88	86	91	93	89	89
Michael	91	89	83	95	80	91	81	96	93	93

21. What is the lowest score Jessica can receive on her next test in order to have a 90 test average?

22. What score does Michael need to make to get an A average?

23. Find the means.

24. Find the interquartile range for each person.

25. According to the different quartiles, whose Q1 is higher?

26. Find the ranges.

27. According to the results of the previous questions, who performed better throughout the year?

28. Find the mean absolute deviation of $\{14, 32\}$.

29. Find the mean absolute deviation of $\{2, 3, 4, 7\}$.

Chapter 10 Test

1. What is the mean of 36, 54, 66, 45, 36, 36, and 63?

   (A) 36
   (B) 45
   (C) 48
   (D) 63

2. What is the median of the following set of data?

   33, 31, 35, 24, 38, 30

   (A) 32
   (B) 31
   (C) 30
   (D) 29

3. Concession stand sales for the first 6 games of the season averaged $400.00. If the total sales of the first 5 games were $320, $540, $230, $450, and $280, what were the total sales for the sixth game?

   (A) $230
   (B) $350
   (C) $364
   (D) $580

4. What is the mean of 12, 23, 8, 26, 37, 11, and 9?

   (A) 12
   (B) 29
   (C) 18
   (D) 19

5. Which of the following sets of numbers has a median of 42?

   (A) {60, 42, 37, 22, 19}
   (B) {16, 28, 42, 48}
   (C) {42, 64, 20}
   (D) {12, 42, 40, 50}

6. The scores on the math quiz were 94, 73, 87, 81, 82, 62, 55, 60. What is the upper quartile value of these scores?

   (A) 77
   (B) 80
   (C) 84.1
   (D) 84.5

7. The cafeteria worker surveyed students in the lunch line about what was their favorite kind of pizza? The workers decide to survey the first 150 students that come into the cafeteria on Monday morning. Which of these **best** describes why this type of sampling may give biased results?

   (A) Students are randomly selected.
   (B) The sample size is too small.
   (C) Not all students have the same chance to complete the survey.
   (D) Every other student entering the cafeteria should be surveyed.

8. There are 72 students in the school orchestra this school year. They have been invited to do a concert for the President of the United States in Washington, which will be televised nation wide. The catch is the concert will be on Homecoming weekend. The music director wants to survey the orchestra about whether or not to accept the invitation. How many orchestra members should he survey to get a representative sample?

   (A) 12
   (B) 24
   (C) 36
   (D) 42

Use the figure for questions 9–10.

Bob	7	4	6	8	7	5	3
Ann	6	5	6	7	9	5	4

9. Find the means. Who produced more?

   (A) Ann, 6
   (B) Ann, 5.7
   (C) Bob, 5.7
   (D) Bob, 6

10. What is/are the mode(s) for Ann's numbers?

   (A) 4
   (B) 5
   (C) 6
   (D) 5 and 6

11. What is the mean absolute deviation of $\{2, 2, 3, 4, 14\}$?

   (A) 5
   (B) 3.6
   (C) 2.8
   (D) 3.0

12. Oxford shirt company wants to test how well their new button attachment machine has been working. They decide to test every tenth shirt that comes off the machine for an hour on Tuesday afternoon. Which of the following is a true statement regarding their sampling technique?

   (A) The Oxford shirt company describes a dependent event.
   (B) The Oxford shirt company describes an unbiased event.
   (C) The event is a simple random sample.
   (D) The sampling technique describes a representative sample.

13. A high school principal wants to survey school parents about requiring school uniforms next year. His school has 2700 families. He cuts up a calendar and puts all the dates from this year in a box. He then has his secretary mix up the papers and draw out 70 dates. He then sends out a survey to the parents of the students that were born on the dates that were selected. Which of the following is a true statement regarding his sampling technique?

   (A) The sampling technique describes a dependent sample.
   (B) The sampling technique describes a biased sample.
   (C) The event described is a simple random sample.
   (D) The sampling technique describes an unrepresentative sample.

Chapter 11
Complex Numbers

This chapter covers the following Georgia Performance Standards:

MA1N	Number and Operations	MA1N1a
		MA1N1b
		MA1N1c
		MA1N1d

Complex numbers are usually written in the form $a + bi$, where a and b are real numbers and i is defined as $\sqrt{-1}$. Because $\sqrt{-1}$ does not exist in the set of real numbers, i is referred to as the imaginary unit.

When talking about a complex number, $a + bi$, the real number a is called the real part, and the real number b is called the imaginary part.

If the real part, a, is zero, then the complex number $a + bi$ is just bi, so it is imaginary.

If the real part, b, is zero, then the complex number $a + bi$ is just a, so it is real.

Example 1:   What is the real part of the complex number $9 + 16i$?

Solution: The complex number $9 + 16i$ is written in the form $a + bi$. Here $a = 9$ and $b = 16$. The real part of the complex number $9 + 16i$ is 9.

Example 2:   What is the imaginary part of the complex number $23 - 6i$?

Solution: The complex number $23 - 6i$ is written in the form $a + bi$. Here $a = 23$ and $b = -6$. The imaginary part of the complex number $a + bi$ is b. The imaginary part of the complex number $23 - 6i$ is -6.

Name the real part of each of the following complex numbers.

1. $-\frac{4}{5} - 3i$ 3. $20 - 11i$ 5. $15i$ 7. $12 + 5i$

2. $7 + 2i$ 4. $\frac{2}{9}$ 6. $-13 + \frac{i}{4}$ 8. $\frac{-9 + 2i}{25}$

Name the imaginary part of each of the following complex numbers.

9. $4 - i$ 11. $5 + \frac{2}{3}i$ 13. 18 15. $51 - 2i$

10. $\frac{6i}{5}$ 12. $-9 + 8i$ 14. $\frac{1 - 3i}{2}$ 16. $14 + i$

11.1 Imaginary Numbers

The square root of a negative number is an imaginary number. You know that $\sqrt{-1} = i$. Therefore, $i^2 = -1$.

Where n is some natural number $(1, 2, 3...)$, then $\sqrt{-n} = \sqrt{(-1) \times n} = \sqrt{-1} \times \sqrt{n} = i\sqrt{n}$. To remove the negative number from under the radical, just take i out. Remember $\sqrt{-n} = i\sqrt{n}$.

Example 3: Simplify: $\sqrt{-450}$

Step 1: Factor -450 and rewrite: $\sqrt{-450} = \sqrt{225 \times 2 \times (-1)}$.

Step 2: By root laws, $\sqrt{225 \times 2 \times (-1)} = \sqrt{225} \times \sqrt{2} \times \sqrt{-1}$.

Step 3: Since $\sqrt{225} = 15$ and $\sqrt{-1} = i$, we have $\sqrt{225} \times \sqrt{2} \times \sqrt{-1} = 15 \times \sqrt{2} \times i$.

Step 4: Write in standard form as $15i\sqrt{2}$.

Example 4: Multiply: $5i \times 2i$

Step 1: Using the basic rules of multiplication, we know $5i \times 2i = 5 \times i \times 2 \times i$.

Step 2: Use the commutative property of multiplication, $5 \times i \times 2 \times i = 5 \times 2 \times i \times i$.

Step 3: Simplify: $5 \times 2 \times i \times i = 10 \times i^2$.

Step 4: Since $i^2 = -1$, then $10 \times i^2$ can be simplified to $10 \times -1 = -10$.

Find the square root of each of the following numbers.

1. -8

2. $-\dfrac{4}{49}$

3. -441

4. $-\dfrac{81}{16}$

5. -44

6. -0.0121

7. -144

8. -64

Use what you know about the imaginary number i to solve the following problems.

9. $-3i + \sqrt{-3}$

10. $7i - 8i$

11. $\sqrt{-4} \times \sqrt{-9}$

12. $2i \times (-4i)$

13. $\left(\sqrt{-16}\right) \div (2i)$

14. $14i + i$

15. $\sqrt{-25} - 3i$

16. $(12i) \div (3i)$

11.2   Adding and Subtracting Complex Numbers

Complex numbers, written as $a + bi$ or $c + di$ may be added and subtracted. The real parts are added or subtracted together and the imaginary parts are added or subtracted together. So, where a, c are the real parts of two complex numbers and b, d are the corresponding imaginary parts.

$$(a + bi) + (c + di) = (a + c) + (b + d)\, i$$

Example 5: What is $(6 + i) - (5 - 7i)$?

Solution: Collect the real parts and the imaginary parts and do the arithmetic.
$(6 + i) - (5 - 7i) = 6 + i - 5 + 7i = (6 - 5) + (i + 7i) = 1 + 8i$

Example 6: What is $(10 + 3i) + (-7 - 6i) + (18 - 5i)$?

Step 1: Group the real and imaginary parts.
$(10 + 3i) + (-7 - 6i) + (18 - 5i) = (10 - 7 + 18) + (3 - 6 - 5)\, i$

Step 2: Add or subtract the real and imaginary parts separately.
$(10 - 7 + 18) + (3 - 6 - 5)\, i = 21 - 8i$

Example 7: z and v are complex numbers. $z = -2 + 3i$ and $v = 3 - 2i$. Compute $z - v$.

Solution: $z - v = (-2 + 3i) - (3 - 2i) = (-2 - 3) + (3i + 2i) = -5 + 5i$

Add.

1. $(5 + 2i) + (-3 - 11i)$
2. $(1 - 5i) + \left(4 + \frac{4i}{3}\right) + (20 - 7i)$
3. $(12 + 2i) + \left(\frac{4}{5} - i\right)$
4. $(3 + 8i) + (4 + 9i) + (2 - 3i)$
5. $(-13 + 4i) + \left(9 - \frac{i}{5}\right)$
6. $22 + (3 - 7i) + (-1 + 20i)$

7. $(16 - 10i) + (-3 - 4i) + 19i$
8. $(2 - 6i) + \left(5 + \frac{9i}{2}\right)$
9. $(30 + 7i) + (-23 - 15i) + (8 + 6i)$
10. $(-4 + i) + (21 - 18i) + 10$
11. $\left(\frac{9}{10} - 17i\right) + (9 + 13i)$
12. $(12 + 2i) + (2 - 12i) + 23i$

Subtract.

13. $(17 + 10i) - (6 + 12i)$
14. $(11 - i) - \left(2 + \frac{2}{11}i\right)$
15. $(3 - 2i) - (-4 + 5i) - (2 - 16i)$
16. $\left(9 + \frac{i}{8}\right) - (20 - 22i)$
17. $(40 + 4i) - (2 + 13i) - i$
18. $(5 - 4i) - (4 + 3i) - (18 + 3i)$

19. $(7 - i) - (10 + 35i) - 8$
20. $(12 + 8i) - (5 - 8i) - (1 + i)$
21. $(6 + 3i) - (5 + 4i) - (11 - i)$
22. $\left(13 - \frac{7i}{6}\right) - \left(31 + \frac{6i}{7}\right)$
23. $(2 + 20i) - (-1 + 6i) - (3 + 14i)$
24. $(-11 - 5i) - (5 - 11i) - 11i$

11.3 Multiplying Complex Numbers

Multiplying two complex numbers, $a+bi$ and $c+di$, should remind you of the FOIL (First Outside Inside Last) method for multiplying two binomials like $(x+2)(x+3) = x^2 + 2x + 3x + 6 = x^2 + 5x + 6$. Generally, multiplying two complex numbers is the same:

$$(a+bi)(c+di) = ac + adi + bci + bdi^2$$

Simplify:

Remember that $i^2 = \left(\sqrt{-1}\right)^2 = -1$, so $ac + adi + bci + bdi^2 = ac + (ad + bc)i - bd$

Example 8:   Multiply $1 + 2i$ and $-4 + 3i$.

Step 1:   Use the FOIL method to multiply:
$(1 + 2i)(-4 + 3i) = (1)(-4) + (1)(3i) + (2i)(-4) + (2i)(3i)$

Step 2:   Simplify:
$(1)(-4) + (1)(3i) + (2i)(-4) + (2i)(3i) = -4 + 3i - 8i + 6i^2 =$
$-4 - 5i + 6i^2 = -4 - 5i + 6(-1)$

Step 3:   Combine like terms:
$-4 - 5i + 6(-1) = -10 - 5i$

Example 9:   What is $-7i \times -4i$?

Solution: $-7i \times -4i = 28i^2 = -28$

Example 10:   Compute $(4 - 8i)(6 + 2i)$.

Solution: Use the FOIL method and add the results.
$(4 - 8i)(6 + 2i) = 24 + 8i - 48i - 16i^2 = 40 - 40i$

Multiply.

1. $(4 + 7i) \times (1 - i)$

2. $(5 - 2i) \times (6 + 3i) \times 2$

3. $(8 + 4i) \times (2 + 5i) \times (4 + i)$

4. $(10 - i) \times (2 - 3i) \times 12i$

5. $(25 + 7i) \times (25 - 7i)$

6. $(3 - i)^3$

7. $\left(\frac{1}{4} + 4i\right) \times (4 - 8i)$

8. $(7 + 5i) \times (7 + 5i)$

9. $(1 - 8i) \times (1 - 8i) \times (1 + 8i)$

10. $(6 - i) \times (7 - i) \times i$

11. $(11 - 9i) \times (11 + 9i)$

12. $(3 + 4i) \times (1 + 10i) \times 3$

11.4   Dividing Complex Numbers

There are three steps to remember when dividing complex numbers.

1. Write the complex number as a fraction.

2. Multiply the numerator and denominator of the fraction by the complex conjugate of the denominator. The **complex conjugate** of $a + bi$ is $a - bi$.

3. Simplify.

Example 11:   What is $(6 + 3i) \div (2 + i)$?

Step 1:   Write the expression as a fraction:
$$\frac{6 + 3i}{2 + i}$$

Step 2:   Multiply the top (numerator) and the bottom (denominator) by $2 - i$, the complex conjugate of the denominator.
$$\frac{6 + 3i}{2 + i} \times \frac{2 - i}{2 - i}$$

Step 3:   Simplify.
$$\frac{6 + 3i}{2 + i} \times \frac{2 - i}{2 - i} = \frac{12 - 6i + 6i - 3i^2}{4 - 2i + 2i - i^2} = \frac{12 - 3(-1)}{4 - (-1)} = \frac{15}{5} = 3$$

Example 12:   What is $(5 - 4i) \div (7 - 8i)$?

Step 1:   Write the expression as a fraction, $\dfrac{5 - 4i}{7 - 8i}$.

Step 2:   Multiply by $\dfrac{7 + 8i}{7 + 8i}$.

Step 3:   Simplify:
$$\frac{5 - 4i}{7 - 8i} \times \frac{7 + 8i}{7 + 8i} = \frac{35 + 40i - 28i - 32i^2}{49 + 56i - 56i - 64i^2} = \frac{35 + 12i - 32(-1)}{49 - 64(-1)}$$
$$= \frac{67 + 12i}{113} = \frac{67}{113} + \frac{12}{113}i.$$

Divide.

1. $(7 + i) \div (4 - 3i)$

2. $25 \div (-1 + 5i)$

3. $(10 - 2i) \div (3 + 6i)$

4. $(-9 - 4i) \div (9 - 4i)$

5. $(5 + 3i) \div (-2 + i)$

6. $(8 - 7i) \div (1 - 8i)$

7. $(-11 - i) \div (7 + 3i)$

8. $(20 + i) \div 9i$

9. $(3 - 8i) \div (-8 + 3i)$

10. $(6 + 2i) \div (10 + i)$

11. $(-1 + 5i) \div (1 - 5i)$

12. $(15 + 2i) \div (3 - 7i)$

11.5   Simplify Complex Numbers

The order of operations for expressions with complex numbers is the same as the order of operations for real number expressions.

The **absolute value** of $a + bi$ is $\sqrt{a^2 + b^2}$.

Example 13:   Simplify the expression $12 \times [(-1 + 4i)^2 + (2 - 3i)] \div i$.

Step 1:   First, $-1 + 4i$ is squared.

$$(-1 + 4i)^2 = (-1 + 4i)(-1 + 4i) = 1 - 4i - 4i + 16i^2 = 1 - 8i + 16(-1) = -15 - 8i$$

Step 2:   Plug the result into the original expression.

$$12 \times [(-15 - 8i) + (2 - 3i)] \div i$$

Step 3:   Add $-15 - 8i$ and $2 - 3i$:

$$(-15 - 8i) + (2 - 3i) = -15 - 8i + 2 - 3i = (-15 + 2) + (-8 - 3)i = -13 - 11i$$

Step 4:   Plug the result from step 3 into equation in step 2:

$$12 \times (-13 - 11i) \div i.$$

Step 5:   Multiply first:

$$12 \times (-13 - 11i) = (12)(-13) + (12)(-11i) = -156 - 132i$$

The equation in step 4 can now be written as $\dfrac{-156 - 132i}{i}$.

Step 6:   Finally, $-156 - 132i$ is divided by i as follows:

$$\frac{-156 - 132i}{i} = \frac{(-156 - 132i)(-i)}{(i)(-i)} = \frac{(-156)(-i) + (-132i)(-i)}{(i)(-i)}$$

$$= \frac{156i + 132i^2}{-i^2} = \frac{156i + 132(-1)}{-(-1)} = \frac{156i - 132}{1} = -132 + 156i$$

Simplify each of the following expressions.

1. $(-10 + i) + (5 - 4i) \times (1 + 13i)$

2. $(4 - 5i) - (7 + 2i) \div (2 - i)^2$

3. $((9 - 3i) + (3 + 9i)) \times (6i + (3 + i))$

4. $((17 - 2i) + (-15 + 4i))^2$

5. $((8 + 5i)^2 - (33 + 75i))^2$

6. $(2 + 11i) \div ((1 - 7i) - (8 - 8i))$

7. $\dfrac{3 + i}{3 - i} + \dfrac{2 + i}{4 + 3i}$

8. $((39 - 36i) - (4 - 5i)^2)^2$

9. What is the absolute value of $5 - 2i$?

10. What is the absolute value of $-3 + 9i$?

Chapter 11 Review

Name the real part of each of the following complex numbers.

1. $-38 + 17i$

2. $\dfrac{8}{5} - \dfrac{13}{5}i$

Name the imaginary part of each of the following complex numbers.

3. $11 - 16i$

4. $225 + 725i$

Use what you know about the imaginary number i to solve the following problems.

5. $-5i + 18i$

6. $\sqrt{-12} \times \sqrt{3}$

Find the square root of each of the following numbers.

7. -64

8. -361

Add.

9. $(-14 - 3i) + (7 + 9i)$

10. $\left(\frac{3}{10} - 20i\right) + \left(-8 + \frac{1}{5}i\right)$

11. $(4 - 15i) + (12 + 19i)$

12. $(-21 + 20i) + (32 - 5i)$

Subtract.

13. $(10 - 4i) - (42 + 7i)$

14. $(-13 + 23i) - (18 - 6i)$

15. $(76 - 52i) - (43 + 27i)$

16. $\left(26 + \frac{4}{9}i\right) - \left(31 - \frac{2}{3}i\right)$

Multiply.

17. $(1 - 6i) \times (5 + 5i)$

18. $(4 + 2i) \times (10 - 15i)$

19. $(8 + i) \times (-4 - 3i)$

20. $(13 - 4i) \times (13 + 4i)$

Divide.

21. $(5 + 4i) \div (-3 - 2i)$

22. $(14 - 7i) \div (7 + i)$

23. $(6 + 3i) \div (-8 - 4i)$

24. $(-2 + 8i) \div (-10 + 2i)$

25. $104 \div (10 - 2i)$

26. $89 \div (-5 + 8i)$

Chapter 11 Test

1. What is $\dfrac{-9-2i}{3-i} + \dfrac{2+7i}{1-4i}$?

(A) $-\dfrac{137}{34} - \dfrac{21}{34}i$

(B) $-\dfrac{137}{34} + \dfrac{21}{34}i$

(C) $\dfrac{137}{34} - \dfrac{21}{34}i$

(D) $\dfrac{137}{34} + \dfrac{21}{34}i$

2. Which of the following statements is true?

(A) Every real number is a complex number with an imaginary part of -1.

(B) Every real number is a complex number with an imaginary part of 0.

(C) Every real number is a complex number with an imaginary part of 1.

(D) Every complex number is a real number with an imaginary part of 1.

3. What is the complex conjugate of $\frac{1}{10} - \frac{3}{10}i$?

(A) $-\frac{1}{10} - \frac{3}{10}i$

(B) $-\frac{1}{10} + \frac{3}{10}i$

(C) $\frac{1}{10} - \frac{3}{10}i$

(D) $\frac{1}{10} + \frac{3}{10}i$

4. What is $(8 - 18i) - (-3 - 13i)$?

(A) $5 - 31i$

(B) $5 - 5i$

(C) $11 - 31i$

(D) $11 - 5i$

5. What is $(4 - i) - \left(8 + \frac{1}{2}i\right) \times (2 + 6i)^2$?

(A) $-272 - 177i$

(B) $-264 - 175i$

(C) $272 - 177i$

(D) $264 - 175i$

6. What is $(2 - 9i) \div (-4 + 6i)$?

(A) $-\dfrac{31}{26} - \dfrac{6}{13}i$

(B) $-\dfrac{31}{26} + \dfrac{6}{13}i$

(C) $\dfrac{31}{26} - \dfrac{6}{13}i$

(D) $\dfrac{31}{26} + \dfrac{6}{13}i$

7. What is $\left(-\frac{5}{6} - \frac{1}{3}i\right) + \left(\frac{1}{2} + \frac{1}{6}i\right)$?

(A) $-\frac{1}{3} - \frac{1}{6}i$

(B) $-\frac{1}{3} + \frac{1}{6}i$

(C) $\frac{1}{3} - \frac{1}{6}i$

(D) $\frac{1}{3} + \frac{1}{6}i$

8. What is the imaginary part of the complex number $\dfrac{76}{3} - \dfrac{32}{3}i$?

(A) i

(B) $-\dfrac{32}{3}$

(C) $\dfrac{32}{3}$

(D) $\dfrac{76}{3}$

9. When multiplying two complex numbers that both have a real part and an imaginary part not equal to 0, which of the following statements is true regarding the result?

(A) It will always have a real part equal to 0.

(B) It will always have an imaginary part equal to 0.

(C) It will sometimes have an imaginary part equal to 0.

(D) It will never have an imaginary part equal to 0.

Chapter 12
More Functions

This chapter covers the following Georgia Performance Standards:

MA1A	Algebra	MA1A1h
		MA1A1i

12.1   Function Symmetry

In Mathematics, functions can be defined as symmetrical with respect to the y-axis or the origin. To test equations for symmetry, it is helpful to remember the following:

$f(-x) = f(x)$ means the function is symmetrical with respect to the y-axis.
Being symmetrical with the y-axis means the function is **even**.

$f(-x) = -f(x)$ means the function is symmetrical with respect to the origin.
Being symmetrical with the origin means the function is **odd**.

$f(-x) \neq -f(x)$ or $f(x)$ then the function is not symmetrical.

Example 1:   Test the following function for symmetry. $f(x) = x^4 + x^2 + 3$

Step 1:   First, we need to substitute $-x$ in the function for x.
$f(-x) = (-x)^4 + (-x)^2 + 3$

Step 2:   Carry out the operations.
$f(-x) = x^4 + x^2 + 3$

Step 3:   Since $f(-x) = x^4 + x^2 + 3$ and $f(x) = x^4 + x^2 + 3$, then $f(-x) = f(x)$.
This means $f(x) = x^4 + x^2 + 3$ is symmetrical with respect to the y-axis.
The function $f(x) = x^4 + x^2 + 3$ is even.

Example 2:   Test the following function for symmetry. $f(x) = x^5 + x^3 + x$

Step 1:   First, we need to substitute $-x$ in the function for x.
$$f(-x) = (-x)^5 + (-x)^3 + (-x)$$

Step 2:   Carry out the operations.
$$f(-x) = -x^5 - x^3 - x$$

Step 3:   Factor out a negative. $f(-x) = -x^5 - x^3 - x = -(x^5 + x^3 + x)$. Since $f(-x) = -(x^5 + x^3 + x)$ and $-f(x) = -(x^5 + x^3 + x)$, then $f(-x) = -f(x)$. This means $f(x) = x^5 + x^3 + x$ is symmetrical with respect to the origin. The function $f(x) = x^5 + x^3 + x$ is odd.

Example 3:   Test the following function for symmetry. $f(x) = x^7 + x^4 - x^2$

Step 1:   First, we need to substitute $-x$ in the function for x.
$$f(-x) = (-x)^7 + (-x)^4 - (-x)^2$$

Step 2:   Carry out the operations.
$$f(-x) = -x^7 + x^4 - x^2$$

Step 3:   In this case, we see that $f(-x) \neq -f(x)$ or $f(x)$, which means this function is neither even nor odd (no symmetry).

Determine whether the function is even, odd, or neither.

1. $f(x) = x^3 + x^2 + x + 1$

2. $f(x) = 2x^3 - 4x$

3. $f(x) = 7x^2 - 11$

4. $f(x) = 8x^2 + x^4$

5. $f(x) = 4x^3 + x$

6. $f(x) = 6x^3 + 2x^2 - x$

7. $f(x) = x^5 + x^3 + 11$

8. $f(x) = x^7 - x^4 + x^2 - 11$

12.2   Symmetry of Graphs of Functions

Just like the symmetry of functions was determined algebraically, it can also be determined by looking at the graph. If you are unable to tell just by looking at the graph, you can check your answers using the following

y-axis (even)	If (a, b) is on the graph, so is $(-a, b)$.
origin (odd)	If (a, b) is on the graph, so is $(-a, -b)$.
x-axis	If (a, b) is on the graph, so is $(a, -b)$.

Example 4:   Determine if the graph of $y = x^2$ is symmetrical. If it is, is it odd, even, or neither?

Step 1:   Graph the function.

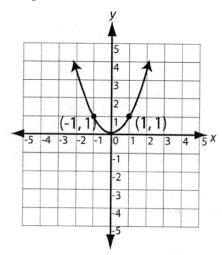

Looking at the graph we see that the graph is symmetrical about the y-axis. In other words, whatever the graph shows on one side of the y-axis, it is the same on the other side of the y-axis.

Step 2:   We check using the points on the graph. We will use the point $(1, 1)$ as it is a solution to the equation.
According to the table above, if $(1, 1)$ is a solution to a graph that is symmetrical about the y-axis, then $(-1, 1)$ must also be a solution. Is it? $y = (-1)^2 = 1$. Yes, it is.
We also know since the graph is symmetrical about the y-axis that the graph is even.

Example 5:   Determine if the graph of $x = y^2$ is symmetrical? If it is, is it odd, even, or neither?

Step 1:   Graph the function.

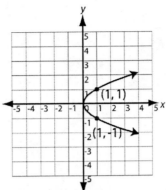

Looking at the graph we see that the graph is symmetrical about the x-axis. In other words, whatever the graph shows on one side of the x-axis, it is the same on the other side of the x-axis.

Step 2:   We check using the points on the graph. We will use the point $(1, 1)$ as it is a solution to the equation. If $(1, 1)$ is a solution to a graph that is symmetrical about the x-axis, then $(1, -1)$ must also be a solution. Is it? Yes.
We also know since the graph is symmetrical about the x-axis that the graph is neither even or odd.

Example 6:   Determine if the graph of $y = \dfrac{1}{x}$ is symmetrical? If it is, is it odd, even, or neither?

Step 1:   Graph the function.

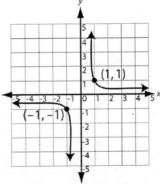

Looking at the graph we see that the graph is symmetrical about the origin. In other words, if you rotate one part of the graph $180°$ around the origin, it will be the exact same as the other half of the graph.

Step 2:   We check using the points on the graph. We will use the point $(1, 1)$ as it is a solution to the equation. If $(1, 1)$ is a solution to a graph that is symmetrical about the origin, then $(-1, -1)$ must also be a solution. Is it? Yes.
We also know since the graph is symmetrical about the origin that the graph is odd.

Tell whether the graph is symmetrical with respect to the x-axis, y-axis, the origin, or neither. Also, tell whether the graph is odd, even, or neither.

1.

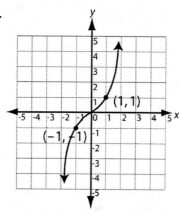

4.

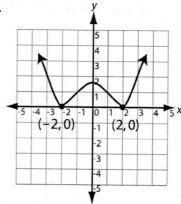

2.

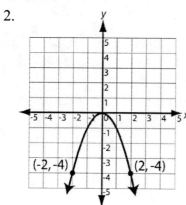

5.

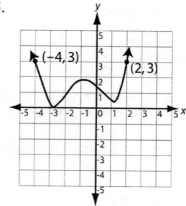

3.

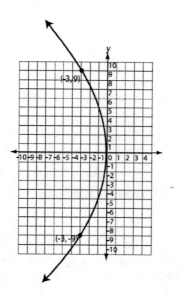

6.

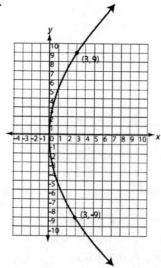

12.3   Finding Common Solutions for Intersecting Lines

When two lines intersect, they share exactly one point in common.

Example 7: $3x + 4y = 20$ and $2y - 4x = 12$

Put each equation in slope-intercept form.

$$3x + 4y = 20 \qquad\qquad 2y - 4x = 12$$
$$4y = -3x + 20 \qquad\qquad 2y = 4x + 12$$
$$y = -\tfrac{3}{4}x + 5 \qquad\qquad y = 2x + 6$$

slope-intercept form

Straight lines with different slopes are **intersecting lines**. Look at the graphs of the lines on the same Cartesian plane.

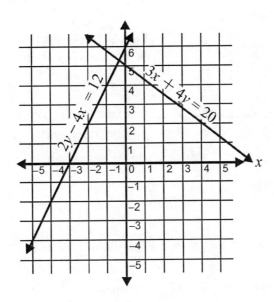

You can see from looking at the graph that the intersecting lines share one point in common.

12.4   Solutions of Equations

An intersection point is a point where two functions meet. To find the intersection point, you must set the equations equal to each other, $f(x) = g(x)$.

Example 8:   Find the intersection point(s) of $f(x) = x + 5$ and $g(x) = 2x + 6$.

Step 1:   Find the value of x in the intersection point by setting the two equations equal to each other.

$$
\begin{aligned}
x + 5 &= 2x + 6 && \text{Set the equations equal to each other.} \\
5 &= x + 6 && \text{Subtract } x \text{ from both sides and simplify.} \\
-1 &= x && \text{Subtract 6 from both sides and simplify.}
\end{aligned}
$$

Step 2:   We use the value of x to find the intersection point by substituting the x-value back into one of the original equations.

$$
\begin{aligned}
f(x) &= x + 5 && \text{Choose an equation.} \\
f(-1) &= -1 + 5 && \text{Substitute the } x\text{-value into the equation.} \\
f(-1) &= 4 && \text{Simplify.}
\end{aligned}
$$

The intersection point is $(-1, 4)$.

Example 9:   Find the intersection point(s) of $f(x) = x^2$ and $g(x) = -3x + 4$.

Step 1:   Find the value of x in the intersection point by setting the two equations equal to each other.

$$
\begin{aligned}
x^2 &= -3x + 4 && \text{Set the equations equal to each other.} \\
x^2 + 3x - 4 &= 0 && \text{Move all terms to one side the equation.} \\
(x + 4)(x - 1) &= 0 && \text{Factor.} \\
x = -4 \text{ or } x &= 1 && \text{Solve for } x.
\end{aligned}
$$

Step 2:   We use the values of x to find the intersection points by substituting the x-values back into one of the original equations.

$$
\begin{aligned}
f(-4) &= (-4)^2 && \text{Substitute one } x\text{-value into an original equation.} \\
f(-4) &= 16 && \text{Solve.}
\end{aligned}
$$

$$
\begin{aligned}
f(1) &= 1^2 && \text{Substitute the other } x\text{-value to find the 2nd intersection point.} \\
f(1) &= 1 && \text{Solve.}
\end{aligned}
$$

The intersection points are $(-4, 16)$ and $(1, 1)$.

Find the intersection points of the pairs of functions.

1. $f(x) = 2x + 3$ and $g(x) = -\frac{1}{2}x + 7$

2. $f(x) = 2x - 3$ and $g(x) = x + 4$

3. $f(x) = 5x - 1$ and $g(x) = 2x + 8$

4. $f(x) = x^2 + 6$ and $g(x) = -5x$

5. $f(x) = x^2 - x$ and $g(x) = 4x - 6$

6. $f(x) = x^2 - 3x$ and $g(x) = -4x + 6$

7. $f(x) = x^2 + 4x$ and $g(x) = 7x + 18$

8. $f(x) = 4x + 4$ and $g(x) = 16x + 4$

Chapter 12 Review

Solve the following problems about function symmetry.

1. True or False. If for a function $f(x)$, $f(-x) = -f(x)$ is true, then the function is odd.

2. True or False. If for a function $f(x)$, $f(-x) = f(x)$ is true, then the function is even.

3. Fill in the blank. When all exponents in a function are ____, the function will be even.

4. Is the function $f(x) = x^6 + x^4 + 7$ even, odd, or neither?

5. Is the function $f(x) = x^5 + x^3 + 2$ even, odd, or neither?

6. Fill in the blank. When all exponents in a function are ____, the function will be odd.

7. Is the function $f(x) = x^4 - x + 2$ even, odd, or neither?

State whether the graphs are symmetrical with respect to the x-axis, y-axis, origin, or neither.

8.

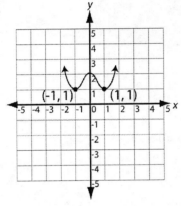

9.

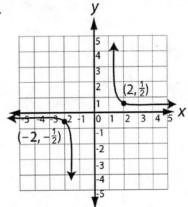

Find the intersection points of the pairs of functions.

10. $f(x) = x^2 - 7x - 8$ and $g(x) = 7x + 7$

11. $f(x) = 7x - 8$ and $g(x) = 5x + 2$

12. $f(x) = x^2 + 2x + 1$ and $g(x) = -2x^2 - 3x - 1$

Chapter 12 Test

1. What is the intersection point of the following functions?

$f(x) = 4x - 8$ and $g(x) = 2x$

(A) $(-4, -8)$
(B) $(4, 8)$
(C) $(-1, -2)$
(D) $(1, 2)$

2. Is the following function even, odd, neither, or not enough information?
$f(x) = x^{10} + x^6 + x^2 + 4$

(A) even
(B) odd
(C) neither
(D) not enough information

3. Is the following function even, odd, neither, or not enough information?
$f(x) = x^5 + x^4 + x^2 + x + 4$

(A) even
(B) odd
(C) neither
(D) not enough information

4. Is the following function even, odd, neither, or not enough information?
$f(x) = 2x^3 - 7x$

(A) even
(B) odd
(C) neither
(D) not enough information

5. Is the following function symmetrical?
$f(x) = 6x^2 - 13$

(A) yes, about the y-axis
(B) yes, about the x-axis
(C) yes, about the origin
(D) no

6. Is the following function symmetrical?
$f(x) = 7x^5 - 4x - 1$

(A) yes, about the y-axis
(B) yes, about the x-axis
(C) yes, about the origin
(D) no

7. Is the following graph symmetrical?

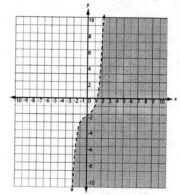

(A) yes, about the y-axis
(B) yes, about the x-axis
(C) yes, about the origin
(D) no

8. How many times do $f(x) = x + 2$ and $g(x) = 2x + 4$ intersect?

(A) 0
(B) 1
(C) 2
(D) 3

9. How many times do $f(x) = x^2$ and $g(x) = -x^2$ intersect?

(A) 0
(B) 1
(C) 2
(D) 3

10. What is the intersection point of $f(x) = -x$ and $g(x) = 2x + 1$?

(A) $\left(\frac{1}{3}, -\frac{1}{3}\right)$

(B) $\left(-\frac{1}{3}, 1\right)$

(C) $\left(-\frac{1}{3}, \frac{1}{3}\right)$

(D) $\left(-\frac{1}{3}, \frac{2}{3}\right)$

11. What is the intersection point of $f(x) = 2x + 1$ and $g(x) = x + 5$?

(A) $(2, 5)$

(B) $\left(\frac{3}{4}, \frac{5}{2}\right)$

(C) $(3, 8)$

(D) $(4, 9)$

12. Where do $f(x) = x^2 - 2x$ and $g(x) = 3x - 6$ intersect?

(A) $(2, 0)$ and $(3, 3)$

(B) $(2, 3)$ and $(3, 0)$

(C) $(2, 1)$ and $(3, 1)$

(D) $(2, 3)$ and $(3, 2)$

Chapter 13
Roots

This chapter covers the following Georgia Performance Standards:

| MA1A | Algebra | MA1A2a |
| | | MA1A2b |

13.1   Square Root

Just as working with exponents is related to multiplication, finding square roots is related to division. In fact, the sign for finding the square root of a number looks similar to a division sign. The best way to learn about square roots is to look at examples.

Examples: This is a square root problem: $\sqrt{64}$
It is asking, "What is the square root of 64?"
It means, "What number multiplied by itself equals 64?"
The answer is 8. $8 \times 8 = 64$.

Find the square roots of the following numbers.

1. $\sqrt{49}$ 4. $\sqrt{16}$ 7. $\sqrt{100}$ 10. $\sqrt{36}$ 13. $\sqrt{64}$

2. $\sqrt{81}$ 5. $\sqrt{121}$ 8. $\sqrt{289}$ 11. $\sqrt{4}$ 14. $\sqrt{9}$

3. $\sqrt{25}$ 6. $\sqrt{625}$ 9. $\sqrt{196}$ 12. $\sqrt{900}$ 15. $\sqrt{144}$

13.2   Simplifying Square Roots

Square roots can sometimes be simplified even if the number under the square root is not a perfect square. One of the rules of roots is that if a and b are two positive real numbers, then it is always true that $\sqrt{a \cdot b} = \sqrt{a} \cdot \sqrt{b}$. You can use this rule to simplify square roots.

Example 1: $\sqrt{100} = \sqrt{4 \cdot 25} = \sqrt{4} \cdot \sqrt{25} = 2 \cdot 5 = 10$

Example 2: $\sqrt{200} = \sqrt{100 \cdot 2} = 10\sqrt{2}$ ⟵ Means 10 multiplied by the square root of 2

Example 3: $\sqrt{160} = \sqrt{10 \cdot 16} = 4\sqrt{10}$

Simplify.

1. $\sqrt{98}$ 3. $\sqrt{50}$ 5. $\sqrt{8}$ 7. $\sqrt{48}$ 9. $\sqrt{54}$ 11. $\sqrt{72}$ 13. $\sqrt{90}$ 15. $\sqrt{18}$

2. $\sqrt{600}$ 4. $\sqrt{27}$ 6. $\sqrt{63}$ 8. $\sqrt{75}$ 10. $\sqrt{40}$ 12. $\sqrt{80}$ 14. $\sqrt{175}$ 16. $\sqrt{20}$

13.3   Extracting Monomial Roots

When finding the roots of monomial expressions, you must first divide the monomial expression into separate parts. Then, simplify each part of the expression.

Note: To find the square root of any variable raised to a positive exponent, simply divide the exponent by 2. For example, $\sqrt{y^{10}} = y^5$.

Example 4: $\sqrt{25x^6y^4z^6}$

    Step 1:   Break each component apart. $\left(\sqrt{25}\right)\left(\sqrt{x^6}\right)\left(\sqrt{y^4}\right)\left(\sqrt{z^6}\right)$

    Step 2:   Solve for each component. $\left(\sqrt{25} = 5\right)\left(\sqrt{x^6} = x^3\right)\left(\sqrt{y^4} = y^2\right)\left(\sqrt{z^6} = z^3\right)$

    Step 3:   Recombine the simplified expressions. $(5)\left(x^3\right)\left(y^2\right)\left(z^3\right) = 5x^3y^2z^3$

Simplify the problems below.

1. $\sqrt{4a^2b^4c^8}$

2. $\sqrt{49h^{24}i^6j^4}$

3. $\sqrt{121p^{20}q^{24}r^6}$

4. $\sqrt{36a^{28}b^{10}c^6}$

5. $\sqrt{144t^{44}u^{30}v^2}$

6. $\sqrt{36k^6l^{26}m^{20}}$

7. $\sqrt{25s^6t^{24}u^{32}}$

8. $\sqrt{81x^6y^{20}z^{44}}$

9. $\sqrt{49a^6b^4c^8}$

10. $\sqrt{169u^{12}v^{24}w^{28}}$

11. $\sqrt{64x^{44}y^{20}z^6}$

12. $\sqrt{4d^{30}e^{54}f^{10}}$

13. $\sqrt{f^4g^6h^{26}}$

14. $\sqrt{900l^{50}m^{26}n^4}$

15. $\sqrt{400g^{40}h^{26}i^{36}}$

16. $\sqrt{25a^{54}b^6c^{46}}$

17. $\sqrt{16j^{24}k^8l^{20}}$

18. $\sqrt{9q^4r^{20}s^{34}}$

13.4   Monomial Roots with Remainders

Monomial roots which are not easily simplified under the square root symbol will also sometimes be encountered. Powers may be raised to odd numbers. In addition, the coefficients may not be perfect squares. Follow the example below to understand how to simplify these types of problems.

Example 5:   Simplify $\sqrt{40x^7y^{11}z^{23}}$

    Step 1:   Begin by simplifying the coefficient. $\sqrt{40} = \left(\sqrt{4}\right)\left(\sqrt{10}\right)$, $\sqrt{4} = 2$, so $\sqrt{40} = 2\sqrt{10}$

    Step 2:   Simplify the variable with exponents.

$$\sqrt{x^7} = \left(\sqrt{x^6}\right)\left(\sqrt{x}\right), \sqrt{x^6} = x^3, \text{ so } \sqrt{x^7} = x^3\sqrt{x}$$

$$\sqrt{y^{11}} = \left(\sqrt{y^{10}}\right)\left(\sqrt{y}\right), \sqrt{y^{10}} = y^5, \text{ so } \sqrt{y^{11}} = y^5\sqrt{y}$$

$$\sqrt{z^{23}} = \left(\sqrt{z^{22}}\right)\left(\sqrt{z}\right), \sqrt{z^{22}} = z^{11}, \text{ so } \sqrt{z^{23}} = z^{11}\sqrt{z}$$

    Step 3:   Recombine the simplified expressions. $2x^3y^5z^{11}\sqrt{10xyz}$

Simplify the following square root expressions.

1. $\sqrt{57d^{25}e^{27}f^{22}}$

2. $\sqrt{140h^{26}i^{20}j^9}$

3. $\sqrt{27x^{44}y^{42}z^9}$

4. $\sqrt{75p^{22}q^8r^{21}}$

5. $\sqrt{48k^{47}l^{27}m^3}$

6. $\sqrt{75s^{23}t^7u^{28}}$

7. $\sqrt{63a^8b^{27}c^{42}}$

8. $\sqrt{20p^3q^{44}r^{29}}$

9. $\sqrt{80a^{220}b^{20}c^{27}}$

10. $\sqrt{64m^8n^3p^{22}}$

11. $\sqrt{88r^{27}s^{22}t^{23}}$

12. $\sqrt{40g^{42}h^{25}j^{28}}$

13. $\sqrt{90v^3w^{20}x^{24}}$

14. $\sqrt{50d^7e^9f^{23}}$

15. $\sqrt{45x^{28}y^6z^{23}}$

16. $\sqrt{32a^6b^{23}c^7}$

17. $\sqrt{74j^{24}k^{27}m^7}$

18. $\sqrt{20q^{24}r^{27}s^7}$

13.5   Adding and Subtracting Roots

You can add and subtract terms with square roots only if the number under the square root sign, the radical, is the same.

Example 6: $2\sqrt{2} + 3\sqrt{2} = 5\sqrt{2}$

Example 7: $12\sqrt{7} - 3\sqrt{7} - 9\sqrt{7}$

Or, look at the following examples where you can simplify the square roots and then add or subtract.

Example 8: $2\sqrt{25} + \sqrt{36}$

   Step 1:   Simplify. You know that $\sqrt{25} = 5$, and $\sqrt{36} = 6$ so the problem simplifies to $2(5) + 6$

   Step 2:   Solve: $2(5) + 6 = 10 + 6 = 16$

Example 9: $2\sqrt{72} - 3\sqrt{2}$

   Step 1:   Simplify what you know. $\sqrt{72} = \sqrt{36 \cdot 2} = 6\sqrt{2}$

   Step 2:   Substitute $6\sqrt{2}$ for $\sqrt{72}$ simplify.
   $2(6)\sqrt{2} - 3\sqrt{2} = 12\sqrt{2} - 3\sqrt{2} = 9\sqrt{2}$

Simplify the following addition and subtraction problems.

1. $3\sqrt{5} + 9\sqrt{5}$

2. $3\sqrt{25} + 4\sqrt{16}$

3. $4\sqrt{8} + 2\sqrt{2}$

4. $3\sqrt{32} - 2\sqrt{2}$

5. $\sqrt{25} - \sqrt{49}$

6. $2\sqrt{5} + 4\sqrt{20}$

7. $5\sqrt{8} - 3\sqrt{72}$

8. $\sqrt{27} + 3\sqrt{27}$

9. $3\sqrt{20} - 4\sqrt{45}$

10. $4\sqrt{45} - \sqrt{75}$

11. $2\sqrt{28} + 2\sqrt{7}$

12. $\sqrt{64} + \sqrt{81}$

13. $5\sqrt{54} - 2\sqrt{24}$

14. $\sqrt{32} + 2\sqrt{50}$

15. $2\sqrt{7} + 4\sqrt{63}$

16. $8\sqrt{2} + \sqrt{8}$

17. $2\sqrt{8} - 4\sqrt{32}$

18. $\sqrt{36} + \sqrt{100}$

19. $\sqrt{9} + \sqrt{25}$

20. $\sqrt{64} - \sqrt{36}$

21. $\sqrt{75} + \sqrt{108}$

22. $\sqrt{81} + \sqrt{100}$

23. $\sqrt{192} - \sqrt{75}$

24. $3\sqrt{5} + \sqrt{245}$

13.6   Multiplying Roots

You can also multiply square roots. To multiply square roots, you just multiply the numbers under the radical and then simplify. Look at the examples below.

Example 10: $\sqrt{2} \times \sqrt{6}$

    Step 1: $\sqrt{2} \times \sqrt{6} = \sqrt{2 \times 6} = \sqrt{12}$ Multiply the numbers under the radical.

    Step 2: $\sqrt{12} = \sqrt{4 \times 3} = 2\sqrt{3}$ Simplify.

Example 11: $3\sqrt{3} \times 5\sqrt{6}$

    Step 1: $(3 \times 5)\sqrt{3 \times 6} = 15\sqrt{18}$ Multiply the numbers in front of the square root, and multiply the numbers under the radical.

    Step 2: $15\sqrt{18} = 15\sqrt{2 \times 9}$ Simplify.
               $15 \times 3\sqrt{2} = 45\sqrt{2}$

Example 12: $\sqrt{14} \times \sqrt{42}$ For this more complicated multiplication problem, use the rule of roots that you learned on page 174, $\sqrt{a \cdot b} = \sqrt{a} \cdot \sqrt{b}$.

    Step 1: $\sqrt{14} = \sqrt{7} \times \sqrt{2}$ and   Instead of multiplying 14 by 42, divide these
           $\sqrt{42} = \sqrt{2} \times \sqrt{3} \times \sqrt{7}$ numbers into their roots.

           $\sqrt{14} \times \sqrt{42} = \sqrt{7} \times \sqrt{2} \times \sqrt{2} \times \sqrt{3} \times \sqrt{7}$

    Step 2:   Since you know that $\sqrt{7} \times \sqrt{7} = 7$ and $\sqrt{2} \times \sqrt{2} = 2$, the problem simplifies to
           $(7 \times 2)\sqrt{3} = 14\sqrt{3}$

Simplify the following multiplication problems.

1. $\sqrt{5} \times \sqrt{7}$

2. $\sqrt{32} \times \sqrt{2}$

3. $\sqrt{10} \times \sqrt{14}$

4. $2\sqrt{3} \times 3\sqrt{6}$

5. $4\sqrt{2} \times 2\sqrt{10}$

6. $\sqrt{5} \times 3\sqrt{15}$

7. $\sqrt{45} \times \sqrt{27}$

8. $5\sqrt{21} \times \sqrt{7}$

9. $\sqrt{42} \times \sqrt{21}$

10. $4\sqrt{3} \times 2\sqrt{12}$

11. $\sqrt{56} \times \sqrt{24}$

12. $\sqrt{11} \times 2\sqrt{33}$

13. $\sqrt{13} \times \sqrt{26}$

14. $2\sqrt{2} \times 5\sqrt{5}$

15. $\sqrt{6} \times \sqrt{12}$

13.7 Dividing Roots

When dividing a number or a square root by another square root, you cannot leave the square root sign in the denominator (the bottom number) of a fraction. You must simplify the problem so that the square root is not in the denominator. This is also called rationalizing the denominator. Look at the examples below.

Example 13: $\dfrac{\sqrt{2}}{\sqrt{5}}$

Step 1: $\dfrac{\sqrt{2}}{\sqrt{5}} \times \dfrac{\sqrt{5}}{\sqrt{5}}$ ⟵ The fraction $\frac{\sqrt{5}}{\sqrt{5}}$ is equal to 1, and multiplying by 1 does not change the value of a number.

Step 2: $\dfrac{\sqrt{2 \times 5}}{5} = \dfrac{\sqrt{10}}{5}$ Multiply and simplify. Since $\sqrt{5} \times \sqrt{5}$ equals 5, you no longer have a square root in the denominator.

Example 14: $\dfrac{6\sqrt{2}}{2\sqrt{10}}$ In this problem, the numbers outside of the square root will also simplify.

Step 1: $\dfrac{6}{2} = 3$ so you have $\dfrac{3\sqrt{2}}{\sqrt{10}}$

Step 2: $\dfrac{3\sqrt{2}}{\sqrt{10}} \times \dfrac{\sqrt{10}}{\sqrt{10}} = \dfrac{3\sqrt{2 \times 10}}{10} = \dfrac{3\sqrt{20}}{10}$

Step 3: $\dfrac{3\sqrt{20}}{10}$ will further simplify because $\sqrt{20} = 2\sqrt{5}$, so you then have $\dfrac{3 \times 2\sqrt{5}}{10}$ which reduces to $\dfrac{3\sqrt{5}}{5}$.

Simplify the following division problems.

1. $\dfrac{9\sqrt{3}}{\sqrt{5}}$

2. $\dfrac{16}{\sqrt{8}}$

3. $\dfrac{24\sqrt{10}}{12\sqrt{3}}$

4. $\dfrac{\sqrt{121}}{\sqrt{6}}$

5. $\dfrac{\sqrt{40}}{\sqrt{90}}$

6. $\dfrac{33\sqrt{15}}{11\sqrt{2}}$

7. $\dfrac{\sqrt{32}}{\sqrt{12}}$

8. $\dfrac{\sqrt{11}}{\sqrt{5}}$

9. $\dfrac{\sqrt{2}}{\sqrt{6}}$

10. $\dfrac{2\sqrt{7}}{\sqrt{14}}$

11. $\dfrac{5\sqrt{2}}{4\sqrt{8}}$

12. $\dfrac{4\sqrt{21}}{7\sqrt{7}}$

13. $\dfrac{9\sqrt{22}}{2\sqrt{2}}$

14. $\dfrac{\sqrt{35}}{2\sqrt{14}}$

15. $\dfrac{\sqrt{40}}{\sqrt{15}}$

16. $\dfrac{\sqrt{3}}{\sqrt{12}}$

Chapter 13 Review

Simplify the following square root expressions.

1. $\sqrt{50}$

2. $\sqrt{44}$

3. $\sqrt{12}$

4. $\sqrt{18}$

5. $\sqrt{8}$

6. $\sqrt{48}$

7. $\sqrt{75}$

8. $\sqrt{200}$

9. $\sqrt{32}$

10. $\sqrt{20}$

11. $\sqrt{63}$

12. $\sqrt{80}$

13. $\sqrt{36r^6 s^8 t^2}$

14. $\sqrt{40g^3 h^6 j^7}$

15. $\sqrt{18m^5 n^3 p^7}$

16. $\sqrt{75a^7 b^2 c^9}$

Simplify the following square root problems.

17. $5\sqrt{27} + 7\sqrt{3}$

18. $\sqrt{40} - \sqrt{10}$

19. $\sqrt{64} + \sqrt{81}$

20. $8\sqrt{50} - 3\sqrt{32}$

21. $14\sqrt{5} + 8\sqrt{80}$

22. $\sqrt{63} \times \sqrt{28}$

23. $\dfrac{\sqrt{56}}{\sqrt{35}}$

24. $\sqrt{8} \times \sqrt{50}$

25. $\dfrac{\sqrt{20}}{\sqrt{45}}$

26. $5\sqrt{40} \times 3\sqrt{20}$

27. $2\sqrt{48} - \sqrt{12}$

28. $\dfrac{2\sqrt{5}}{\sqrt{30}}$

29. $\dfrac{3\sqrt{22}}{2\sqrt{3}}$

30. $\sqrt{72} \times 3\sqrt{27}$

Chapter 13 Test

1. Simplify: $\sqrt{135}$

   (A) $3\sqrt{15}$

   (B) $\sqrt{72}$

   (C) $9\sqrt{15}$

   (D) $\sqrt{9} \times \sqrt{15}$

2. Express $\dfrac{\sqrt{20}}{\sqrt{35}}$ in simplest form.

   (A) $\dfrac{2\sqrt{7}}{7}$

   (B) $\dfrac{2}{\sqrt{7}}$

   (C) $\dfrac{2\sqrt{5}}{\sqrt{7}}$

   (D) $\dfrac{4}{7}$

3. Simplify: $\sqrt{44} \cdot 2\sqrt{33}$

   (A) $2\sqrt{77}$

   (B) $44\sqrt{3}$

   (C) $22\sqrt{7}$

   (D) $22\sqrt{12}$

4. Simplify: $\sqrt{12x^3y^4z^7}$

   (A) $2xyz^3\sqrt{3xyz}$

   (B) $xy^2z^3\sqrt{12xz}$

   (C) $2xy^2z^3\sqrt{3xz}$

   (D) $2xy^2z^3\sqrt{3}$

5. Simplify: $\sqrt{45} \times \sqrt{27}$

   (A) $3\sqrt{15}$

   (B) $\sqrt{72}$

   (C) $9\sqrt{15}$

   (D) $\sqrt{9} \times \sqrt{15}$

6. Simplify: $\dfrac{3\sqrt{12}}{2\sqrt{3}}$

   (A) 3

   (B) $\dfrac{3\sqrt{4}}{2}$

   (C) $\dfrac{6}{\sqrt{6}}$

   (D) $3\sqrt{3}$

7. Simplify the following by rationalizing the denominator.

   $$\dfrac{\sqrt{3}}{\sqrt{15}}$$

   (A) $\dfrac{\sqrt{5}}{5}$

   (B) $\dfrac{1}{\sqrt{5}}$

   (C) $\dfrac{\sqrt{45}}{15}$

   (D) $\dfrac{3\sqrt{5}}{15}$

8. Simplify: $\sqrt{64f^6g^7h^5}$

   (A) $8f^3g^3h^2\sqrt{gh}$
   (B) $8f^3g^3h^2$
   (C) $f^3g^3h^2\sqrt{64gh}$
   (D) $8f^3g^3h^2\sqrt{g}$

Chapter 14
Polynomials

This chapter covers the following Georgia Performance Standards:

| MA1A | Algebra | MA1A2c |
| | | MA1A2f |

Polynomials are algebraic expressions which include **monomials** containing one term, **binomials** which contain two terms, and **trinomials**, which contain three terms. Expressions with more than three terms are called **polynomials**. **Terms** are separated by plus and minus signs.

EXAMPLES

Monomials	Binomials	Trinomials	Polynomials
$5f$	$5t + 20$	$x^2 + 4x + 3$	$x^3 - 3x^4 + 3x - 20$
$3x^3$	$20 - 8g$	$7x^4 - 6x - 2$	$p^5 + 4p^3 + p^4 + 20p - 7$
$5g^4$	$7x^4 + 8x$	$y^5 + 27y^4 + 200$	
4	$6x^3 - 9x$		

14.1   Adding and Subtracting Monomials

Two **monomials** are added or subtracted as long as the **variable and its exponent** are the **same**. This is called combining like terms. Use the same rules you used for adding and subtracting integers

Example 1: $5x + 7x = 12x$ $\begin{array}{r} 3x^5 \\ -9x^5 \\ \hline -6x^5 \end{array}$ $4x^4 - 20x^4 = -16x^4$ $\begin{array}{r} 7y \\ +4y \\ \hline 11y \end{array}$ $6y^3 - 7y^3 = -y^3$

Remember: When the integer in front of the variable is "1", the one is usually not written. $1x^4$ is the same as x^4, and $-1x$ is the same as $-x$.

Add or subtract the following monomials.

1. $4x^4 + 7x^4 =$

2. $7t + 9t =$

3. $20y^3 - 4y^3 =$

4. $6g - 9g =$

5. $8y^4 + 9y^4 =$

6. $s^7 + s^7 =$

7. $-4x - 5x =$

8. $5w^4 - w^4 =$

9. $z^5 + 20z^5 =$

10. $-k + 4k =$

11. $3x^4 - 7x^4 =$

12. $20t + 4t =$

13. $-8v^3 + 20v^3 =$

14. $-4x^3 + x^3 =$

15. $20y^5 - 7y^5 =$

16. y^5
    $+4y^5$
    ___

18. $9t^4$
    $+8t^4$
    ___

20. $7w^4$
    $+9w^4$
    ___

22. $-7z$
    $+20z$
    ___

24. $8t^3$
    $-6t^3$
    ___

17. $5x^3$
    $-20x^3$
    ___

19. $-4y$
    $-5y$
    ___

21. $22t^3$
    $-5t^3$
    ___

23. $5w^7$
    $+w^7$
    ___

25. $3x$
    $+9x$
    ___

14.2 Adding Polynomials

When adding **polynomials,** make sure the exponents and variables are the same on the terms you are combining. The easiest way is to put the terms in columns with **like exponents** under each other. Each column is added as a separate problem. Fill in the blank spots with zeros if it helps you keep the columns straight. You never carry to the next column when adding polynomials.

Example 2:   Add $3x^4 + 25$ and $7x^4 + 4x$

$$\begin{array}{r} 3x^4 + 0x + 25 \\ (+)\,7x^4 + 4x + 0 \\ \hline 10x^4 + 4x + 25 \end{array}$$

Example 3: $(5x^3 - 4x) + (-x^3 - 5)$

$$\begin{array}{r} 5x^3 - 4x + 0 \\ (+)\, -x^3 + 0x - 5 \\ \hline 4x^3 - 4x - 5 \end{array}$$

Add the following polynomials.

1. $y^4 + 3y + 4$ and $4y^4 + 5$

2. $(7y^4 + 5y - 6) + (4y^4 - 7y + 9)$

3. $-4x^4 + 7x^3 + 5x - 2$ and $3x^4 - x + 4$

4. $-p + 5$ and $7p^4 - 4p + 4$

5. $(w - 4) + (w^4 + 4)$

6. $5t^4 - 7t - 8$ and $9t + 4$

7. $t^5 + t + 9$ and $4t^3 + 5t - 5$

8. $(s^4 + 3s^3 - 4) + (-4s^3 + 5)$

9. $(-v^4 + 8v - 9) + (5v^3 - 6v + 5)$

10. $6m^4 - 4m + 20$ and $m^4 - m - 9$

11. $-x + 5$ and $3x^4 + x - 4$

12. $(9t^4 + 3t) + (-8t^4 - t + 5)$

13. $(3p^5 + 4p^4 - 2) + (-7p^4 - p + 9)$

14. $20s^4 + 24s^3 + 4s$ and $s^4 + s^3 + s$

15. $(-20b^4 + 8b + 4) + (-b^4 + 6b + 20)$

16. $27c^4 - 22c + 7$ and $-8c^4 + 3c - 20$

17. $4c^4 + 7c^3 + 3$ and $5c^4 + 4c^3 + 2$

18. $3x^4 + -25x^3 + 27$ and $8x^3 - 24$

19. $(-x^4 + 4x - 5) + (3x^4 - 3)$

20. $(y^4 - 22y + 20) + (-23y^4 + 7y - 5)$

21. $3d^7 - 5d^3 + 8$ and $4d^5 - 4d^3 - 4$

22. $(6t^7 - t^3 + 28) + (5t^7 + 8t^3)$

23. $5p^4 - 9p + 20$ and $-p^4 - 3p - 7$

24. $40b^3 + 27b$ and $-5b^4 - 7b + 25$

25. $(-4w + 22) + (w^3 + w - 5)$

26. $(47z^4 + 23z + 9) + (z^4 - 4z - 20)$

14.3   Subtracting Polynomials

When you subtract polynomials, it is important to remember to change all the signs in the subtracted polynomial (the subtrahend) and then add.

Example 4: $(5y^4 + 9y + 20) - (4y^4 + 6y - 5)$

Step 1:   Copy the subtraction problem into vertical form. Make sure you line up the terms with like exponents under each other just like you did for adding polynomials.

$$\begin{array}{r} 5y^4 + 9y + 20 \\ (-)\,4y^4 + 6y - 5 \\ \hline \end{array}$$

Step 2:   Change the subtraction sign to addition and all the signs of the subtracted polynomial to the opposite sign. The bottom polynomial in the problem becomes $-4y^4 - 6y + 5$.

Step 3:   Add:
$$\begin{array}{r} 5y^4 + 9y + 20 \\ (+)-4y^4 - 6y + 5 \\ \hline y^4 + 3y + 25 \end{array}$$

Subtract the following polynomials.

1. $(4x^4 + 7x + 4) - (x^4 + 3x + 2)$

2. $(9y - 5) - (5y + 3)$

3. $(-5t^4 + 22t^3 + 3) - (5t^4 - t^3 - 7)$

4. $(-3w^4 + 20w - 7) - (-7w^4 - 7)$

5. $(6a^7 - a^3 + a) - (8a^7 + a^4 - 3a)$

6. $(25c^5 + 40c^4 + 20) - (8c^5 + 7c^4 + 24)$

7. $(7x^4 - 20x) - (-8x^4 + 5x + 9)$

8. $(-9y^4 + 24y^3 - 20) - (3y^3 + y + 20)$

9. $(-3h^4 - 8h + 8) - (7h^4 + 5h + 20)$

10. $(20k^3 - 9) - (k^4 - 5k^3 + 7)$

11. $(x^4 - 7x + 20) - (6x^4 - 7x + 8)$

12. $(24p^4 + 5p) - (20p - 4)$

13. $(-4m - 9) - (6m + 4)$

14. $(4y^4 + 23y^3 - 9y) - (5y^4 + 4y^3 - 8y)$

15. $(8g + 3) - (g^4 + 5g - 7)$

16. $(-9w^3 + 5w) - (-5w^4 - 20w^3 - w)$

17. $(x^4 + 24x^3 - 20) - (4x^4 + 3x^3 + 2)$

18. $(4a^4 + 4a + 4) - (-a^4 + 3a + 3)$

19. $(c + 220) - (3c^4 - 8c + 4)$

20. $(-6v^4 + 24v) - (3v^4 + 4v + 6)$

21. $(3b^4 + 5b^3 + 7) - (8b^3 - 9)$

22. $(7x^4 + 27x^3 - 5) - (-5x^4 + 5x^3)$

23. $(9y^4 - 4) - (22y^4 - 4y - 3)$

24. $(-z^4 - 7z - 9) - (3z^4 - 7z + 7)$

A subtraction of polynomials problem may be stated in sentence form. Study the examples below.

Example 5:   Subtract $-7x^3 + 5x - 3$ from $3x^3 + 5x^4 - 6x$.

Step 1:   Copy the problem in columns with terms with the same exponent and variable under each other. Notice the second polynomial in the sentence will be the top polynomial of the problem.

$$
\begin{array}{l}
3x^3 + 5x^4 - 6x \\
(-) - 7x^3 \qquad + 5x - 3 \\
\hline
\end{array}
$$

Since this is a subtraction problem, change all the signs of the terms in the bottom polynomial. Then add.

$$
\begin{array}{l}
3x^3 + 5x^4 - 6x \\
(+)\, 7x^3 \qquad - 5x + 3 \\
\hline
10x^3 + 5x^4 - 11x + 3
\end{array}
$$

Example 6:   From $6y^4 + 4$ subtract $5y^4 - 3y + 9$

In a problem phrased like this one, the first polynomial will be on top, and the second will be on bottom. Change the signs on the bottom polynomial and then add.

$$
\begin{array}{l}
6y^4 \qquad + 4 \\
(-)\, 5y^4 - 3y + 9 \\
\hline
\end{array}
\qquad\longrightarrow\qquad
\begin{array}{l}
6y^4 \qquad + 4 \\
(+) - 5y^4 + 3y - 9 \\
\hline
y^4 + 3y - 5
\end{array}
$$

Solve the following subtraction problems.

1. Subtract $3x^4 + 4x - 7$ from $7x^4 + 4$

2. From $7y^3 - 6y + 20$ subtract $9y^3 - 20$

3. From $5m^4 - 5m + 8$ subtract $4m - 3$

4. Subtract $9z^4 + 3z + 4$ from $5z^4 - 8z + 9$

5. Subtract $t^4 + 20t^3 - 7$ from $-t^4 - 4t^3 - 7$

6. Subtract $-8b^3 - 4b + 5$ from $-b^4 + b + 6$

7. From $20y^3 + 40$ subtract $7y^3 - 7$

8. From $25t^4 - 6t - 9$ subtract $5t^4 - 3t + 4$

9. Subtract $3p^4 + p - 4$ from $-8p^4 - 7p + 4$

10. Subtract $x^3 + 9$ from $-4x^4 + 3x^3 + 20$

11. Subtract $24a^4 + 20$ from $-a^4 + a^3 - 2$

12. From $6m^4 + 3m + 2$ subtract $-6m^4 - 3m$

13. From $-3z^4 - 23z^3 - 4$ subtract $-40z^3 + 40$

14. Subtract $20c^4 + 20$ from $9c^4 - 7c + 3$

15. Subtract $b^4 + b - 7$ from $7b^4 - 5b + 7$

16. Subtract $-3x - 5$ from $3x^4 + x + 20$

17. From $27y^4 + 4$ subtract $5y^4 + 3y + 8$

18. Subtract $3g^4 - 7g + 7$ from $20g^4 - 3g - 5$

19. From $-8m^4 - 9m$ subtract $3m^4 + 8$

20. Subtract $x + 2$ from $7x + 7$

21. Subtract $c^4 + c + 4$ from $-c^4 - c - 4$

22. From $6t^4 + 9t^3 - 5t + 4$ subtract $t^3 + 3t$

14.4   Multiplying Monomials

When two monomials have the **same variable**, you can multiply them. Then, add the **exponents** together. If the variable has no exponent, it is understood that the exponent is 1.

Example 7: $5x^5 \times 3x^4 = 15x^9$ $4y \times 7y^4 = 28y^5$

Multiply the following monomials.

1. $6a \times 20a^7$

2. $4x^6 \times 7x^3$

3. $5y^3 \times 3y^4$

4. $20t^4 \times 4t^4$

5. $4p^7 \times 5p^4$

6. $20b^4 \times 9b$

7. $3c^3 \times 3c^3$

8. $4d^9 \times 20d^4$

9. $6k^3 \times 7k^4$

10. $8m^7 \times m$

11. $22z \times 4z^8$

12. $3w^5 \times 6w^7$

13. $5x^5 \times 7x^3$

14. $7n^4 \times 3n^3$

15. $9w^8 \times w$

16. $20s^6 \times 7s^3$

17. $5d^7 \times 5d^7$

18. $7y^4 \times 9y^6$

19. $8t^{20} \times 3t^7$

20. $6p^9 \times 4p^3$

21. $x^3 \times 4x^3$

When problems include negative signs, follow the rules for multiplying integers.

22. $-8s^5 \times 7s^3$

23. $-6a \times -20a^7$

24. $5x \times -x$

25. $-3y^4 \times -y^3$

26. $-7b^4 \times 3b^7$

27. $20c^5 \times -4c$

28. $-5t^3 \times 9t^3$

29. $20d \times -9d^8$

30. $-3g^6 \times -4g^3$

31. $-8s^5 \times 8s^3$

32. $-d^3 \times -4d$

33. $22p \times -4p^7$

34. $-7x^8 \times -3x^3$

35. $9z^5 \times 8z^5$

36. $-5w \times -7w^9$

37. $-7y^5 \times 6y^4$

38. $20x^3 \times -8x^7$

39. $-a^5 \times -a$

40. $-8k^4 \times 3k$

41. $-27t^4 \times -t^5$

42. $3x^9 \times 20x^4$

14.5   Multiplying Monomials with Different Variables

Warning: You cannot add the exponents of variables that are different.

Example 8: $(-5wx)(6w^3x^4)$

To work this problem, first multiply the whole numbers: $-5 \times 6 = -30$. Then multiply the w's: $w \times w^3 = w^4$. Last, multiply the x's: $x \times x^4 = x^5$. The answer is $-30w^4x^5$.

Multiply the following monomials.

1. $(4x^4y^4)(-5xy^3) =$

2. $(20p^3q^5)(4p^4q) =$

3. $(-3t^5v^4)(t^4v) =$

4. $(8w^3z^4)(3wz) =$

5. $(-4st^6)(-9s^4t) =$

6. $(xy^3)(5x^4y^4) =$

7. $(7y^4z)(3y^5z^4) =$

8. $(-3a^4b^4)(-5ab^3) =$

9. $(-7c^3d^4)(4c^5d^7) =$

10. $(20x^5y^4)(3x^3y) =$

11. $(6f^3g^7)(-f^3g) =$

12. $(-5a^3v^5)(9a^5v) =$

13. $(7m^9n^7)(8m^4n^5) =$

14. $(8w^7y^3)(3wy) =$

15. $(4x^5z^4)(-20x^4z^5) =$

16. $(-5a^8c^{20})(4a^4c) =$

17. $(-bd^6)(-b^4d) =$

18. $(3x^5y^4)(20x^3y^3) =$

19. $(20p^4y)(7p^7y^3) =$

20. $(-4a^8x^4)(6ax^4) =$

21. $(9c^5d^3)(-4c^4d^4) =$

Multiplying three monomials works the same way. The first one is done for you.

22. $(3st)(5s^3t^4)(4s^4t^5) = 60s^8t^{10}$

23. $(xy)(x^4y^4)(4x^3y^4) =$

24. $(4a^4b^4)(a^3b^3)(4ab) =$

25. $(5y^4z^5)(4y^3)(4z^4) =$

26. $(7cd^3)(3c^4d^4)(d^4) =$

27. $(4w^4x^3)(3x^5)(4w^3) =$

28. $(a^5d^4)(ad)(a^4d^3) =$

29. $(8x^3t)(4t^5)(x^4t^4) =$

30. $(p^4y^4)(5py)(p^3y^3) =$

31. $(5x^3y)(7xy^3)(4y^5) =$

32. $(9xy^4)(x^4y^3)(4x^4y) =$

33. $(6p^3t)(4t^3)(p^4) =$

34. $(3bc)(b^4c)(5c^3) =$

35. $(4y^5z^7)(4y^6)(y^4z^4) =$

36. $(5p^3r^3)(5r^4)(p^4r) =$

37. $(a^5z^6)(6a^4z^4)(3z^3) =$

38. $(7c^3)(6d^4)(4c^4d) =$

39. $(20s^8t^4)(3st)(s^4t^3) =$

40. $(3a^3b^5)(4b^3)(3a^4) =$

41. $(7wz)(w^3z^3)(3w^4z^3) =$

14.6   Dividing Monomials

When simplifying monomial fractions with exponents, all exponents need to be positive. If there are negative exponents in your answer, put the base with its negative exponent below the fraction line and remove the negative sign. Two variables that are alike should not appear in both the denominator and numerator of a simplified expression. If you have the same variable in both the denominator and numerator of the fraction, the expression is not in its simplest form.

Example 9: $\dfrac{55x^4y^4}{11x^6y^6}$

    Step 1:   Reduce the whole numbers first. $\dfrac{55}{11} = 5$

    Step 2:   Simplify the x's. $\dfrac{x^4}{x^6} = x^{4-6} = x^{-2} = \dfrac{1}{x^2}$

    Step 3:   Simplify the y's. $\dfrac{y^4}{y^6} = y^{4-6} = y^{-2} = \dfrac{1}{y^2}$

    Therefore $\dfrac{55x^4y^4}{11x^6y^6} = \dfrac{5}{x^2y^2}$

Simplify the expressions below. All answers should only have positive exponents.

1. $\dfrac{7xy^3}{x\left(4x^3\right)y^5}$

2. $\dfrac{4a^4b^7}{3a^5b^4}$

3. $\dfrac{8\left(4a^4\right)b^5}{9ab^6}$

4. $\dfrac{26\left(x^4y^5\right)^3}{40xy}$

5. $\dfrac{20a^5b^4}{7a^6b^3}$

6. $\dfrac{7\left(9x^4y^3\right)}{5\left(x^4y^4\right)^4}$

7. $\dfrac{24\left(3a^4\right)b^4}{6a^4b^4}$

8. $\dfrac{\left(6x^3y^5\right)^4}{\left(4x^7y\right)^3}$

9. $\dfrac{27a^4b^3}{3a^7b^6}$

10. $\dfrac{33x^7y^3}{44x^8y^7}$

11. $\dfrac{27\left(4a^6b^8\right)}{42a^3b^6}$

12. $\dfrac{30x^5y^4}{6\left(x^4y^4\right)^4}$

13. $\dfrac{20\left(9ab^7\right)}{40a^4b^4}$

14. $\dfrac{20x^{20}y^8}{57x^7y^3}$

15. $\dfrac{\left(a^5b^8\right)^5}{a^9b^8}$

16. $\dfrac{8\left(x^3y^7\right)}{9x^4y^5}$

17. $\dfrac{20\left(a^3b^5\right)}{5a^7b^4}$

18. $\dfrac{48\left(x^4y^7\right)^4}{42x^3y^5}$

14.7   Multiplying Monomials by Polynomials

Example 10: $-7t\left(4t^4 - 8t + 20\right)$

Step 1:   Multiply $-7t \times 4t^4 = -28t^5$

Step 2:   Multiply $-7t \times -8t = 56t^2$

Step 3:   Multiply $-7t \times 20 = -140t$

Step 4:   Arrange the answers horizontally in order: $-28t^5 + 56t^2 - 140t$

Remove parentheses in the following problems.

1. $3x\left(3x^4 + 5x - 2\right)$

2. $5y\left(y^3 - 8\right)$

3. $8a^4\left(4a^4 + 3a + 4\right)$

4. $-7d^3\left(d^4 - 7d\right)$

5. $4w\left(-5w^4 + 3w - 9\right)$

6. $9p\left(p^3 - 6p + 7\right)$

7. $-20b^4\left(-4b + 7\right)$

8. $4t\left(t^4 - 5t - 20\right)$

9. $20c\left(5c^4 + 3c - 8\right)$

10. $6z\left(4z^5 - 7z^4 - 5\right)$

11. $-20t^4\left(3t^4 + 7t + 6\right)$

12. $c\left(-3c - 7\right)$

13. $3p\left(-p^4 + p^3 - 20\right)$

14. $-k^4\left(4k + 5\right)$

15. $-3\left(5m^4 - 7m + 9\right)$

16. $6x\left(-8x^3 + 20\right)$

17. $-w\left(w^4 - 5w + 8\right)$

18. $4y\left(7y^4 - y\right)$

19. $3d\left(d^7 - 8d^3 + 5\right)$

20. $-7t\left(-5t^4 - 9t + 2\right)$

21. $8\left(4w^4 - 20w + 5\right)$

22. $3y^4\left(y^4 - 22\right)$

23. $v^4\left(v^4 + 3v + 3\right)$

24. $9x\left(4x^3 + 3x + 2\right)$

25. $-7d\left(5d^4 + 8d - 4\right)$

26. $-k^4\left(-3k + 6\right)$

27. $3x\left(-x^4 - 7x + 7\right)$

28. $5z\left(5z^5 - z - 8\right)$

29. $-7y\left(20y^3 - 3\right)$

30. $4b^4\left(8b^4 + 5b + 5\right)$

14.8   Dividing Polynomials by Monomials

Example 11: $\dfrac{-8wx + 6x^2 - 16wx^2}{2wx}$

Step 1:   Rewrite the problem. Divide each term from the top by the denominator, $2wx$.

$$\frac{-8wx}{2wx} + \frac{6x^2}{2wx} + \frac{-16wx^2}{2wx}$$

Step 2:   Simplify each term in the problem. Then combine like terms.

$$-4 + \frac{3x}{w} - 8x$$

Simplify each of the following.

1. $\dfrac{bc^4 - 9bc - 4b^4c^4}{4bc}$

2. $\dfrac{3jk^4 + 24k + 20j^4k}{3jk}$

3. $\dfrac{7x^4y - 9xy^4 + 4y^3}{4xy}$

4. $\dfrac{26st^4 + st - 24s}{5st}$

5. $\dfrac{5wx^4 + 6wx - 24w^3}{4wx}$

6. $\dfrac{cd^4 + 20cd^3 + 26c^4}{4cd}$

7. $\dfrac{y^4z^3 - 4yz - 9z^4}{-4yz^4}$

8. $\dfrac{a^4b + 4ab^4 - 25ab^3}{4a^4}$

9. $\dfrac{pr^4 + 6pr + 9p^4r^4}{4pr^4}$

10. $\dfrac{6xy^4 - 3xy + 29x^4}{-3xy}$

11. $\dfrac{6x^4y + 24xy - 45y^4}{6xy}$

12. $\dfrac{7m^4n - 20mn - 47n^4}{7mn}$

13. $\dfrac{st^4 - 20st - 26s^4t^4}{4st}$

14. $\dfrac{8jk^4 - 25jk - 63j^4}{8jk}$

14.9   Removing Parentheses and Simplifying

In the following problem, you must multiply each term inside the parentheses by the numbers and variables outside the parentheses, and then add the polynomials to simplify the expressions.

Example 12: $9x \left(4x^4 - 7x + 8\right) - 3x \left(5x^4 + 3x - 9\right)$

Step 1:   Multiply to remove the first set of parentheses.

$9x \left(4x^4 - 7x + 8\right) = 36x^5 - 63x^2 + 72x$

Step 2:   Multiply to remove the second set of parentheses.

$-3x \left(5x^4 + 3x - 9\right) = -15x^5 - 9x^2 + 27x$

Step 3:   Copy each polynomial in columns, making sure the terms with the same variable and exponent are under each other. Add to simplify.

$$36x^5 - 63x^2 + 72x$$
$$(+) - 15x^5 - 9x^2 + 27x$$
$$\overline{21x^5 - 72x^2 + 99x}$$

Remove the parentheses and simplify the following problems.

1. $5t \left(t + 8\right) + 7t \left(4t^4 - 5t + 2\right)$

2. $-7y \left(3y^4 - 7y + 3\right) - 6y \left(y^4 - 5y - 5\right)$

3. $-3 \left(3x^4 + 5x\right) + 7x \left(x^4 + 3x + 4\right)$

4. $4b \left(7b^4 - 9b - 2\right) - 3b \left(5b + 3\right)$

5. $9d^4 \left(3d + 5\right) - 8d \left(3d^4 + 5d + 7\right)$

6. $7a \left(3a^4 + 3a + 2\right) - \left(-4a^4 + 7a - 5\right)$

7. $3m \left(m + 8\right) + 9 \left(5m^4 + m + 5\right)$

8. $5c^4 \left(-6c^4 - 3c + 4\right) - 8c \left(7c^3 + 4c\right)$

9. $-9w \left(-w + 2\right) - 5w \left(3w - 7\right)$

10. $6p \left(4p^4 - 5p - 6\right) + 3p \left(p^4 + 6p + 20\right)$

14.10   Multiplying Two Binomials

When you multiply two binomials such as $(x + 6)(x - 7)$, you must multiply each term in the first binomial by each term in the second binomial. The easiest way is to use the **FOIL** method. If you can remember the word **FOIL**, it can help you keep order when you multiply. The "**F**" stands for **first**, "**O**" stands for **outside**, "**I**" stands for **inside**, and "**L**" stands for **last**.

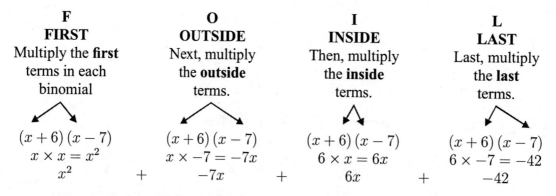

F **FIRST** Multiply the **first** terms in each binomial	**O** **OUTSIDE** Next, multiply the **outside** terms.	**I** **INSIDE** Then, multiply the **inside** terms.	**L** **LAST** Last, multiply the **last** terms.
$(x + 6)(x - 7)$ $x \times x = x^2$ x^2	$(x + 6)(x - 7)$ $x \times -7 = -7x$ $-7x$	$(x + 6)(x - 7)$ $6 \times x = 6x$ $6x$	$(x + 6)(x - 7)$ $6 \times -7 = -42$ -42

Now just combine like terms, $6x - 7x = -x$, and write your answer.
$(x + 6)(x - 7) = x^2 - x - 42$.

Note: It is customary for mathematicians to write polynomials in descending order. That means that the term with the highest-number exponent comes first in a polynomial. The next highest exponent is second and so on. When you use the **FOIL** method, the terms will always be in the customary order. You just need to combine like terms and write your answer.

Multiply the following binomials.

1. $(y - 8)(y + 3)$

2. $(4x + 5)(x + 20)$

3. $(5b - 3)(3b - 5)$

4. $(6g + 4)(g - 20)$

5. $(8k - 7)(-5k - 3)$

6. $(9v - 4)(3v + 5)$

7. $(20p + 4)(5p + 3)$

8. $(3h - 20)(-4h - 7)$

9. $(w - 5)(w - 8)$

10. $(6x + 2)(x - 4)$

11. $(7t + 3)(4t - 2)$

12. $(5y - 20)(5y + 20)$

13. $(a + 6)(3a + 7)$

14. $(3z - 9)(z - 5)$

15. $(7c + 4)(6c + 7)$

16. $(y + 3)(y - 3)$

17. $(4w - 7)(5w + 6)$

18. $(8x + 2)(x - 5)$

19. $(6t - 20)(5t - 5)$

20. $(7b + 6)(6b + 4)$

21. $(4z + 2)(20z + 5)$

22. $(22w - 9)(w + 3)$

23. $(7d - 20)(20d + 20)$

24. $(20g + 4)(g - 4)$

25. $(5p + 8)(4p + 3)$

26. $(m + 7)(m - 7)$

27. $(9b - 9)(4b - 2)$

28. $(z + 3)(3z + 7)$

29. $(8y - 7)(y - 3)$

30. $(20x + 7)(3x - 2)$

31. $(3t + 2)(t + 20)$

32. $(4w - 20)(9w + 8)$

33. $(9s - 4)(s + 5)$

34. $(5k - 2)(9k + 20)$

35. $(h + 24)(h - 4)$

36. $(3x + 8)(8x + 3)$

37. $(4v - 6)(4v + 6)$

38. $(4x + 9)(4x - 3)$

39. $(k - 2)(6k + 24)$

40. $(3w + 22)(4w + 4)$

41. $(9y - 20)(7y - 3)$

42. $(6d + 23)(d - 2)$

43. $(8h + 3)(4h + 5)$

44. $(7n + 20)(7n - 7)$

45. $(6z + 7)(z - 9)$

46. $(5p + 7)(4p - 20)$

47. $(b + 4)(7b + 8)$

48. $(20y - 3)(9y - 8)$

14.11   Perimeter and Area with Algebraic Expressions

You have already calculated the perimeter and area of various shapes with given measurements. You must also understand how to find the perimeter of shapes that are described by algebraic expressions. Study the examples below.

Example 13:   Use the equation $P = 2l + 2w$ to find the perimeter of the following rectangle.

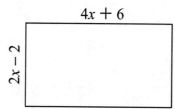

Step 1:   Find $2l$. $2(4x + 6) = 8x + 12$
Step 2:   Find $2w$. $2(2x - 2) = 4x - 4$
Step 3:   Find $2l + 2w$. $12x + 8$

Perimeter $= 12x + 8$

Example 14:   Using the formula $A = lw$, find the area of the rectangle below.

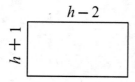

Step 1: $A = (h - 2)(h + 1)$
Step 2: $A = h^2 - 2h + h - 2$
Step 3: $A = h^2 - h - 2$

Area $= h^2 - h - 2$

Example 15:   Find the area of the shaded part in the following figure.

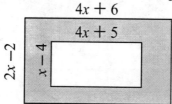

Step 1:   Find the area of the larger rectangle.
$(4x + 6)(2x - 2) = 8x^2 - 8x + 12x - 12 = 8x^2 + 4x - 12$

Step 2:   Find the area of the smaller rectangle.
$(4x + 5)(x - 4) = 4x^2 - 16x + 5x - 20 = 4x^2 - 11x - 20$

Step 3:   Subtract the area of the smaller rectangle from the area of the larger rectangle.

$$\begin{array}{r} 8x^2 + 4x - 12 \\ -\,(4x^2 - 11x - 20) \\ \hline \end{array} \qquad \begin{array}{r} 8x^2 + 4x - 12 \\ -4x^2 + 11x + 20 \quad \leftarrow \text{ Change signs} \\ \hline 4x^2 + 15x + 8 \quad \leftarrow \text{ Area of shaded section} \end{array}$$

Find the perimeter of each of the following rectangles.

1.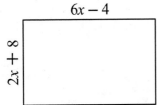

2.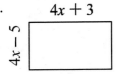

3.

4.

5.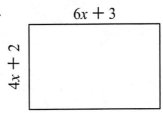

6.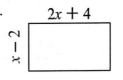

Find the area of each of the following rectangles.

7.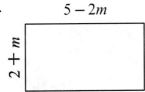

8.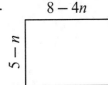

9.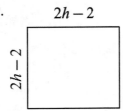

10.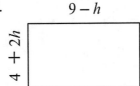

11.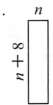

12.

Find the area of the shaded portion of each figure below.

13.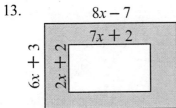

14.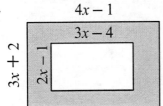

15.

16.

17.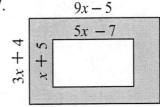

18.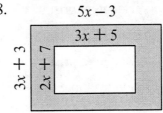

14.12   Volume with Algebraic Expressions

Volume is the amount of space that can be filled by an object expressed in cubic units.

Equations:

Cube	**Rectangular Prism**
$V = s \times s \times s = s^3$	$V = l \times w \times h$

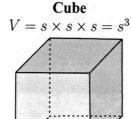

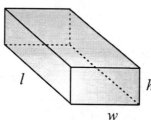

Example 16:   Find the volume of a cube with side length $(x - 2)$.

Step 1:   Write the equation: $V = (x-2)^3 = (x-2) \times (x-2) \times (x-2)$

Step 2:   FOIL the first two factors: $V = (x^2 - 4x - 4)(x-2)$

Step 3:   FOIL again: $V = x^3 - 4x^2 - 4x - 2x^2 + 8x + 8$

Step 4:   Simplify: $V = x^3 - 6x^2 + 4x + 8$

Example 17:   Find the volume of a rectangular prism with height x, length $(x+1)$, and width $(x-7)$.

Step 1:   Substitute values into equation: $V = l \times w \times h = (x+1)(x-7)(x)$

Step 2:   FOIL the first two factors: $V = (x^2 - 7x + x - 7)(x)$

Step 3:   Distribute: $V = x^3 - 7x^2 + x^2 - 7x$

Step 4:   Simplify: $V = x^3 - 6x^2 - 7x$

Find the volume of the figure.

1. a cube with side length $(x + 7)$

2. a cube with side length $(3 + x)$

3. a cube with side length $(x - 9)$

4. Find the volume of the cube in problem number 1 when $x = 2$.

5. a rectangular prism with length $(x + 2)$, width $(x + 3)$, and height (x)

6. a rectangular prism with width $(x - 2)$, length $(7 + x)$, height $(x + 9)$

7. a rectangular prism with length $(x + 1)$, height $(x + 4)$, and width $(12 + x)$

8. Find the volume of the rectangular prism in problem number 6 when $x = 7$.

Chapter 14 Review

Simplify.

1. $3a^4 + 20a^4$

2. $(8x^4y^5)(20xy^7)$

3. $-6z^4(z+3)$

4. $(5b^4)(7b^3)$

5. $8x^4 - 20x^4$

6. $(7p-5)-(3p+4)$

7. $-7t(3t+20)^2$

8. $(3w^3y^4)(5wy^7)$

9. $3(4g+3)^2$

10. $25d^5 - 20d^5$

11. $(8w-5)(w-9)$

12. $27t^4 + 5t^4$

13. $(8c^5)(20c^4)$

14. $(20x+4)(x+7)$

15. $5y(5y^4 - 20y + 4)$

16. $(9a^5b)(4ab^3)(ab)$

17. $(7w^6)(20w^{20})$

18. $9x^3 + 24x^3$

19. $27p^7 - 22p^7$

20. $(3s^5t^4)(5st^3)$

21. $(5d+20)(4d+8)$

22. $5w(-3w^4 + 8w - 7)$

23. $45z^6 - 20z^6$

24. $-8y^3 - 9y^3$

25. $(8x^5)(8x^7)$

26. $28p^4 + 20p^4$

27. $(a^4v)(4av)(a^3v^6)$

28. $5(6y-7)^2$

29. $(3c^4)(6c^9)$

30. $(5x^7y^3)(4xy^3)$

31. Add $4x^4 + 20x$ and $7x^4 - 9x + 4$

32. $5t(6t^4 + 5t - 6) + 9t(3t+3)$

33. Subtract $y^4 + 5y - 6$ from $3y^4 + 8$

34. $4x(5x^4 + 6x - 3) + 5x(x+3)$

35. $(6t-5)-(6t^4 + t - 4)$

36. $(5x+6)+(8x^4 - 4x + 3)$

37. Subtract $7a-4$ from $a+20$

38. $(-4y+5)+(5y-6)$

39. $4t(t+6) - 7t(4t+8)$

40. Add $3c-5$ and $c^4 - 3c - 4$

41. $4b(b-5) - (b^4 + 4b + 2)$

42. $(6k^4 + 7k)+(k^4 + k + 20)$

43. $(q^4r^3)(3qr^4)(4q^5r)$

44. $(7df)(d^5f^4)(4df)$

45. $(8g^4h^3)(g^3h^6)(6gh^3)$

46. $(9v^4x^3)(3v^6x^4)(4v^5x^5)$

47. $(3n^4m^4)(20n^4m)(n^3m^8)$

48. $(22t^4a^4)(5t^3a^9)(4t^6a)$

49. $\dfrac{24(4a^3)b}{3a^4b^{-4}}$

50. $\dfrac{8(g^3h^3)}{5(g^4h)^{-4}}$

51. $\dfrac{26(m^4n^3)^4}{5(m^4n)^{-4}}$

52. $\dfrac{25p^3q^3}{4p^4q}$

53. $\dfrac{9(e^5h^{-4})^{-4}}{36e^4h^7}$

54. $\dfrac{44x^3y^5}{154(x^{-3}y^8)^4}$

Use what you know about volume to solve the following questions.

55. Find the volume of a rectangular prism with length $(x-6)$, width $(9+x)$, and height $(x-11)$.

56. Find the volume of the rectangular prism in question 55 when $x = 3$.

57. Find the volume of a cube with side length $(x+4)$.

58. Find the volume of the cube in question 57 when $x = 4$.

Chapter 14 Test

1. $2x^2 + 5x^2 =$

   (A) $10x^4$
   (B) $7x^4$
   (C) $7x^2$
   (D) $10x^2$

2. $-8m^3 + m^3 =$

   (A) $-8m^6$
   (B) $-8m^9$
   (C) $-9m^6$
   (D) $-7m^3$

3. $(6x^3 + x^2 - 5) + (-3x^3 - 2x^2 + 4) =$

   (A) $3x^3 - x^2 - 1$
   (B) $3x^3 - 3x^2 - 1$
   (C) $3x^3 - 3x^2 - 9$
   (D) $-3x^3 - 3x^2 - 1$

4. $(-7c^2 + 5c + 3) + (-c^2 - 7c + 2) =$

   (A) $-3x^3 - 3x^2 - 1$
   (B) $-8c^2 - 2c + 5$
   (C) $-6c^2 - 12c + 5$
   (D) $-8c^2 - 12c + 5$

5. $(5x^3 - 4x^2 + 5) - (-2x^3 - 3x^2) =$

   (A) $3x^3 + x^2 + 5$
   (B) $3x^3 - 7x^2 + 5$
   (C) $7x^3 - x^2 + 5$
   (D) $7x^3 - 7x^2 + 5$

6. $(-z^3 - 4z^2 - 6) - (3z^3 - 6z + 5) =$

   (A) $-4z^3 - 4z^2 + 6z - 11$
   (B) $-2z^3 - 10z - 1$
   (C) $-4z^3 - 10z^2 - 1$
   (D) $-2z^2 + 2z - 11$

7. $(-7d^5)(-3d^2) =$

   (A) $-21d^7$
   (B) $21d^{10}$
   (C) $21d^7$
   (D) $-21d^{10}$

8. $(-5c^3d)(3c^5d^3)(2cd^4) =$

   (A) $30c^{15}d^8$
   (B) $15c^8d^{12}$
   (C) $-17c^{15}d^{12}$
   (D) $-30c^9d^8$

9. $-11j^2 \times -j^4 =$

   (A) $11j^6$
   (B) $11j^8$
   (C) $-11j^6$
   (D) $-11j^8$

10. $-6m^2(7m^2 + 5m - 6) =$

   (A) $-42m^2 + 30m^3 - 36$
   (B) $-42m^4 - 30m^3 + 36m^2$
   (C) $-13m^4 - m^2 + 36m^2$
   (D) $42m^4 - 30m^3 - 36m^2$

11. $-h^2(-4h + 5) =$

   (A) $-4h^3 - 5h^2$
   (B) $4h^3 - 5h^2$
   (C) $-5h^2 - 5h^2$
   (D) $-5h^3 - 5h^2$

12. $\dfrac{4xy^2 - 6xy + 8x^2y}{2xy} =$

   (A) $2xy - 3 + 4x$
   (B) $2y - 3 + 4xy$
   (C) $2y - 3 + 4x$
   (D) $2xy - 3 + 4x^2$

13. $\dfrac{3cd^3 + 6c^2d - 12cd}{3cd} =$

   (A) $cd + 3c - 4cd$
   (B) $d^2 + 2c - 4cd$
   (C) $cd^2 + 2c - 4cd$
   (D) $d^2 + 2c - 4$

14. $4m(m - 5) + 3m(2m^2 - 6m + 4) =$

   (A) $6m^3 - 14m^2 - 8m$
   (B) $-8m^2 - 8m - 1$
   (C) $7m - 14m^2 - 1$
   (D) $10m^2 - 26m - 20$

15. $2h(3h^2 - 5h - 2) + 4h(h^2 + 6h + 8) =$

   (A) $6h^3 + 19h^2 + 28h$
   (B) $-8m^2 - 8m - 1$
   (C) $7m - 14m^2 - 1$
   (D) $10h^3 + 14h^2 + 28h$

16. Multiply the following binomial and simplify. $(x - 3)(x + 3)$

   (A) $x^2 - 3x + 3x - 9$
   (B) $x^2 - 9$
   (C) $x^2 + 9$
   (D) $x^2 + 6x + 9$

17. Multiply the following binomial and simplify. $(x + 9)(x + 1)$

   (A) $x^2 + 10x + 9$
   (B) $x^2 + 10x + 10$
   (C) $x^2 + 9x + 9$
   (D) $x^2 + 9x + x + 9$

18. Multiply the following binomial and simplify. $(x - 2)^2$

   (A) $x^2 - 4x - 4$
   (B) $x^2 - 2x + 4$
   (C) $x^2 - 2x - 4$
   (D) $x^2 - 4x + 4$

19. The length of the rectangle is 5 units longer than the width. Which expression could be used to represent the area of the rectangle?

   (A) $w^2 + 5w$
   (B) $w^2 + 5$
   (C) $w^2 + 25$
   (D) $w^2 + 10w + 25$

20. The volume of a cube with side length $(x - 3)$ is 8, what is x?

   (A) 3
   (B) 4
   (C) 5
   (D) 11

21. Given the figure with length $(x - 4)$, width $(x - 2)$, and height (x), what is the volume when $x = 7$?

   (A) 100
   (B) 105
   (C) 130
   (D) 115

22. $(x + 4)^2 = ?$

   (A) $x^2 + 4$
   (B) $x^2 + 16$
   (C) $x^2 + 16x + 8$
   (D) $x^2 + 8x + 16$

Chapter 15
Factoring

This chapter covers the following Georgia Performance Standards:

| MA1A | Algebra | MA1A2d |
| | | MA1A2e |

15.1   Finding the Greatest Common Factor of Polynomials

In a multiplication problem, the numbers multiplied together are called **factors**. The answer to a multiplication problem is a called the **product**.

In the multiplication problem $5 \times 4 = 20$, 5 and 4 are factors and 20 is the product.

If we reverse the problem, $20 = 5 \times 4$, we say we have **factored** 20 into 5×4.

In this chapter, we will factor **polynomials**.

Example 1:   Find the greatest common factor of $2y^3 + 6y^2$.

Step 1:   Look at the whole numbers. The greatest common factor of 2 and 6 is 2. Factor the 2 out of each term.

$$2\left(y^3 + 3y^2\right)$$

Step 2:   Look at the remaining terms, $y^3 + 3y^2$. What are the common factors of each term?

$$
\begin{array}{ccccccc}
y^3 & = & y & \times & \boxed{y & \times & y} \\
3y^2 & = & 3 & \times & \boxed{y & \times & y}
\end{array}
\longleftarrow \text{ common factors} = y^2
$$

Step 3:   Factor 2 and y^2 out of each term: $2y^2\left(y + 3\right)$

Check: $2y^2\left(y + 3\right) = 2y^3 + 6y^2$

Factor by finding the greatest common factor in each of the following.

1. $6x^4 + 18x^2$

2. $14y^3 + 7y$

3. $4b^5 + 12b^3$

4. $10a^3 + 5$

5. $2y^3 + 8y^2$

6. $6x^4 - 12x^2$

7. $18y^2 - 12y$

8. $15a^3 - 25a^2$

9. $4x^3 + 16x^2$

10. $6b^2 + 21b^5$

11. $27m^3 + 18m^4$

12. $100x^4 - 25x^3$

13. $4b^4 - 12b^3$

14. $18c^2 + 24c$

15. $20y^3 + 30y^5$

16. $16x^2 - 24x^5$

17. $15a^4 - 25a^2$

18. $24b^3 + 16b^6$

19. $36y^4 + 9y^2$

20. $42x^3 + 49x$

Factoring larger polynomials with 3 or 4 terms works the same way.

Example 2: $4x^5 + 16x^4 + 12x^3 + 8x^2$

Step 1:   Find the greatest common factor of the whole numbers. 4 can be divided evenly into 4, 16, 12, and 8; therefore, 4 is the greatest common factor.

Step 2:   Find the greatest common factor of the variables. x^5, x^4, x^3, and x^2 can be divided by x^2, the lowest power of x in each term.

$$4x^5 + 16x^4 + 12x^3 + 8x^2 = 4x^2 \left(x^3 + 4x^2 + 3x + 2\right)$$

Factor each of the following polynomials.

1. $5a^3 + 15a^2 + 20a$

2. $18y^4 + 6y^3 + 24y^2$

3. $12x^5 + 21x^3 + x^2$

4. $6b^4 + 3b^3 + 15b^2$

5. $14c^3 + 28c^2 + 7c$

6. $15b^4 - 5b^2 + 20b$

7. $t^3 + 3t^2 - 5t$

8. $8a^3 - 4a^2 + 12a$

9. $16b^5 - 12b^4 - 10b^2$

10. $20x^4 + 16x^3 - 24x^2 + 28x$

11. $40b^7 + 30b^5 - 50b^3$

12. $20y^4 - 15y^3 + 30y^2$

13. $4m^5 + 8m^4 + 12m^3 + 6m^2$

14. $16x^5 + 20x^4 - 12x^3 + 24x^2$

15. $18y^4 + 21y^3 - 9y^2$

16. $3n^5 + 9n^3 + 12n^2 + 15n$

17. $4d^6 - 8d^2 + 2d$

18. $10w^2 + 4w + 2$

19. $6t^3 - 3t^2 + 9t$

20. $25p^5 - 10p^3 - 5p^2$

21. $18x^4 + 9x^2 - 36x$

22. $6b^4 - 12b^2 - 6b$

23. $y^3 + 3y^2 - 9y$

24. $10x^5 - 2x^4 + 4x^2$

Example 3:   Find the greatest common factor of $4a^3b^2 - 6a^2b^2 + 2a^4b^3$

Step 1:   The greatest common factor of the whole numbers is 2.

$$4a^3b^2 - 6a^2b^2 + 2a^4b^3 = 2\left(2a^3b^2 - 3a^2b^2 + a^4b^3\right)$$

Step 2:   Find the lowest power of each variable that is in each term. Factor them out of each term. The lowest power of a is a^2. The lowest power of b is b^2.

$$4a^3b^2 - 6a^2b^2 + 2a^4b^3 = 2a^2b^2\left(2a - 3 + a^2b\right)$$

Factor each of the following polynomials.

1. $3a^2b^2 - 6a^3b^4 + 9a^2b^3$

2. $12x^4y^3 + 18x^3y^4 - 24x^3y^3$

3. $20x^2y - 25x^3y^3$

4. $12x^2y - 20x^2y^2 + 16xy^2$

5. $8a^3b + 12a^2b + 20a^2b^3$

6. $36c^4 + 42c^3 + 24c^2 - 18c$

7. $14m^3n^4 - 28m^3n^2 + 42m^2n^3$

8. $16x^4y^2 - 24x^3y^2 + 12x^2y^2 - 8xy^2$

9. $32c^3d^4 - 56c^2d^3 + 64c^3d^2$

10. $21a^4b^3 + 27a^2b^3 + 15a^3b^2$

11. $4w^3t^2 + 6w^2t - 8wt^2$

12. $5pq^3 - 2p^2q^2 - 9p^3q$

13. $49x^3t^3 + 7xt^2 - 14xt^3$

14. $9cd^4 - 3d^4 - 6c^2d^3$

15. $12a^2b^3 - 14ab + 10ab^2$

16. $25x^4 + 10x - 20x^2$

17. $bx^3 - b^2x^2 + b^3x$

18. $4k^3a^2 + 22ka + 16k^2a^2$

19. $33w^4y^2 - 9w^3y^2 + 24w^2y^2$

20. $18x^3 - 9x^5 + 27x^2$

15.2   Factor By Grouping

Not all polynomials have a common factor in each term. In this case they may sometimes be factored by grouping.

Example 4:   Factor $ab + 4a + 2b + 8$

    Step 1:   Factor an a from the first two terms and a 2 from the last two terms.

        $a(b + 4) + 2(b + 4)$

        Now the polynomial has two terms, $a(b + 4)$ and $2(b + 4)$. Notice that $(b + 4)$ is a factor of each term.

    Step 2:   Factor out the common factor of each term:

        $ab + 4a + 2b + 8 = (b + 4)(a + 2)$.

    Check: Multiply using the FOIL method to check.

        $(b + 4)(a + 2) = ab + 4a + 2b + 8$

Factor the following polynomials by grouping.

1. $xy + 4x + 2y + 8$

2. $cd + 5c + 4d + 20$

3. $xy - 4x + 6y - 24$

4. $ab + 6a + 3b + 18$

5. $ab + 3a - 5b - 15$

6. $xy - 2x + 6x - 12$

7. $cd + 4c + 4d + 16$

8. $mn - 5m + 3n - 15$

9. $ab + 4a + 3b + 12$

10. $xy + 7x - 4y - 28$

11. $ab - 2a + 8b - 16$

12. $cd + 4c - 5d - 20$

13. $mn + 6m - 2n - 12$

14. $xy - 9x - 3y + 27$

15. $bc - 3b + 5c - 15$

16. $ab + a + 7b + 7$

17. $xy + 4y + 2y + 8$

18. $cd + 9c - d - 9$

19. $ab + 2a - 7b - 14$

20. $xy - 6x - 2y + 12$

21. $wz + 6z - 4w - 24$

15.3   Finding the Numbers

The next kind of factoring we will do requires thinking of two numbers with a certain sum and a certain product.

Example 5:   Which two numbers have a sum of 8 and a product of 12? In other words, what pair of numbers would answer both equations?

$$\underline{\hspace{1cm}} + \underline{\hspace{1cm}} = 8 \quad \text{and} \quad \underline{\hspace{1cm}} \times \underline{\hspace{1cm}} = 12$$

You may think $4 + 4 = 8$, but 4×4 does not equal 12.
Or you may think $7 + 1 = 8$, but 7×1 does not equal 12.

$6 + 2 = 8$ and $6 \times 2 = 12$, so 6 and 2 are the pair of numbers that will work in both equations.

For each problem below, find one pair of numbers that will solve both equations.

1. $\underline{\hspace{1.5cm}} + \underline{\hspace{1.5cm}} = 14$ and $\underline{\hspace{1.5cm}} \times \underline{\hspace{1.5cm}} = 40$

2. $\underline{\hspace{1.5cm}} + \underline{\hspace{1.5cm}} = 10$ and $\underline{\hspace{1.5cm}} \times \underline{\hspace{1.5cm}} = 21$

3. $\underline{\hspace{1.5cm}} + \underline{\hspace{1.5cm}} = 18$ and $\underline{\hspace{1.5cm}} \times \underline{\hspace{1.5cm}} = 81$

4. $\underline{\hspace{1.5cm}} + \underline{\hspace{1.5cm}} = 12$ and $\underline{\hspace{1.5cm}} \times \underline{\hspace{1.5cm}} = 20$

5. $\underline{\hspace{1.5cm}} + \underline{\hspace{1.5cm}} = 7$ and $\underline{\hspace{1.5cm}} \times \underline{\hspace{1.5cm}} = 12$

6. $\underline{\hspace{1.5cm}} + \underline{\hspace{1.5cm}} = 8$ and $\underline{\hspace{1.5cm}} \times \underline{\hspace{1.5cm}} = 15$

7. $\underline{\hspace{1.5cm}} + \underline{\hspace{1.5cm}} = 10$ and $\underline{\hspace{1.5cm}} \times \underline{\hspace{1.5cm}} = 25$

8. $\underline{\hspace{1.5cm}} + \underline{\hspace{1.5cm}} = 14$ and $\underline{\hspace{1.5cm}} \times \underline{\hspace{1.5cm}} = 48$

9. $\underline{\hspace{1.5cm}} + \underline{\hspace{1.5cm}} = 12$ and $\underline{\hspace{1.5cm}} \times \underline{\hspace{1.5cm}} = 36$

10. $\underline{\hspace{1.5cm}} + \underline{\hspace{1.5cm}} = 17$ and $\underline{\hspace{1.5cm}} \times \underline{\hspace{1.5cm}} = 72$

11. $\underline{\hspace{1.5cm}} + \underline{\hspace{1.5cm}} = 15$ and $\underline{\hspace{1.5cm}} \times \underline{\hspace{1.5cm}} = 56$

12. $\underline{\hspace{1.5cm}} + \underline{\hspace{1.5cm}} = 9$ and $\underline{\hspace{1.5cm}} \times \underline{\hspace{1.5cm}} = 18$

13. $\underline{\hspace{1.5cm}} + \underline{\hspace{1.5cm}} = 13$ and $\underline{\hspace{1.5cm}} \times \underline{\hspace{1.5cm}} = 40$

14. $\underline{\hspace{1.5cm}} + \underline{\hspace{1.5cm}} = 16$ and $\underline{\hspace{1.5cm}} \times \underline{\hspace{1.5cm}} = 63$

15. $\underline{\hspace{1.5cm}} + \underline{\hspace{1.5cm}} = 10$ and $\underline{\hspace{1.5cm}} \times \underline{\hspace{1.5cm}} = 16$

16. $\underline{\hspace{1.5cm}} + \underline{\hspace{1.5cm}} = 8$ and $\underline{\hspace{1.5cm}} \times \underline{\hspace{1.5cm}} = 16$

17. $\underline{\hspace{1.5cm}} + \underline{\hspace{1.5cm}} = 9$ and $\underline{\hspace{1.5cm}} \times \underline{\hspace{1.5cm}} = 20$

18. $\underline{\hspace{1.5cm}} + \underline{\hspace{1.5cm}} = 13$ and $\underline{\hspace{1.5cm}} \times \underline{\hspace{1.5cm}} = 36$

19. $\underline{\hspace{1.5cm}} + \underline{\hspace{1.5cm}} = 15$ and $\underline{\hspace{1.5cm}} \times \underline{\hspace{1.5cm}} = 50$

20. $\underline{\hspace{1.5cm}} + \underline{\hspace{1.5cm}} = 11$ and $\underline{\hspace{1.5cm}} \times \underline{\hspace{1.5cm}} = 30$

15.4   More Finding the Numbers

Now that you have mastered positive numbers, take up the challenge of finding pairs of negative numbers or pairs where one number is negative and one is positive.

Example 6:   Which two numbers have a sum of -3 and a product of -40? In other words, what pair of numbers would answer both equations?

$$\underline{\hspace{1cm}} + \underline{\hspace{1cm}} = -3 \quad \text{and} \quad \underline{\hspace{1cm}} \times \underline{\hspace{1cm}} = -40$$

It is faster to look at the factors of 40 first. 8 and 5 and 10 and 4 are possibilities. 8 and 5 have a difference of 3, and in fact, $5 + (-8) = -3$ and $5 \times (-8) = -40$. This pair of numbers, 5 and -8, will satisfy both equations.

For each problem below, find one pair of numbers that will solve both equations.

1. $\underline{\hspace{1cm}} + \underline{\hspace{1cm}} = -2$ and $\underline{\hspace{1cm}} \times \underline{\hspace{1cm}} = -35$

2. $\underline{\hspace{1cm}} + \underline{\hspace{1cm}} = 4$ and $\underline{\hspace{1cm}} \times \underline{\hspace{1cm}} = -5$

3. $\underline{\hspace{1cm}} + \underline{\hspace{1cm}} = 4$ and $\underline{\hspace{1cm}} \times \underline{\hspace{1cm}} = -12$

4. $\underline{\hspace{1cm}} + \underline{\hspace{1cm}} = -6$ and $\underline{\hspace{1cm}} \times \underline{\hspace{1cm}} = 8$

5. $\underline{\hspace{1cm}} + \underline{\hspace{1cm}} = 3$ and $\underline{\hspace{1cm}} \times \underline{\hspace{1cm}} = -40$

6. $\underline{\hspace{1cm}} + \underline{\hspace{1cm}} = 10$ and $\underline{\hspace{1cm}} \times \underline{\hspace{1cm}} = -11$

7. $\underline{\hspace{1cm}} + \underline{\hspace{1cm}} = 6$ and $\underline{\hspace{1cm}} \times \underline{\hspace{1cm}} = -27$

8. $\underline{\hspace{1cm}} + \underline{\hspace{1cm}} = 8$ and $\underline{\hspace{1cm}} \times \underline{\hspace{1cm}} = -20$

9. $\underline{\hspace{1cm}} + \underline{\hspace{1cm}} = -5$ and $\underline{\hspace{1cm}} \times \underline{\hspace{1cm}} = -24$

10. $\underline{\hspace{1cm}} + \underline{\hspace{1cm}} = -3$ and $\underline{\hspace{1cm}} \times \underline{\hspace{1cm}} = -28$

11. $\underline{\hspace{1cm}} + \underline{\hspace{1cm}} = -2$ and $\underline{\hspace{1cm}} \times \underline{\hspace{1cm}} = -48$

12. $\underline{\hspace{1cm}} + \underline{\hspace{1cm}} = -1$ and $\underline{\hspace{1cm}} \times \underline{\hspace{1cm}} = -20$

13. $\underline{\hspace{1cm}} + \underline{\hspace{1cm}} = -3$ and $\underline{\hspace{1cm}} \times \underline{\hspace{1cm}} = 2$

14. $\underline{\hspace{1cm}} + \underline{\hspace{1cm}} = 1$ and $\underline{\hspace{1cm}} \times \underline{\hspace{1cm}} = -30$

15. $\underline{\hspace{1cm}} + \underline{\hspace{1cm}} = -7$ and $\underline{\hspace{1cm}} \times \underline{\hspace{1cm}} = 12$

16. $\underline{\hspace{1cm}} + \underline{\hspace{1cm}} = 6$ and $\underline{\hspace{1cm}} \times \underline{\hspace{1cm}} = -16$

17. $\underline{\hspace{1cm}} + \underline{\hspace{1cm}} = 5$ and $\underline{\hspace{1cm}} \times \underline{\hspace{1cm}} = -24$

18. $\underline{\hspace{1cm}} + \underline{\hspace{1cm}} = -4$ and $\underline{\hspace{1cm}} \times \underline{\hspace{1cm}} = 4$

19. $\underline{\hspace{1cm}} + \underline{\hspace{1cm}} = -1$ and $\underline{\hspace{1cm}} \times \underline{\hspace{1cm}} = -42$

20. $\underline{\hspace{1cm}} + \underline{\hspace{1cm}} = -6$ and $\underline{\hspace{1cm}} \times \underline{\hspace{1cm}} = 8$

15.5   Factoring Trinomials

In the chapter on polynomials, you multiplied binomials (two terms) together, and the answer was a trinomial (three terms).

For example, $(x + 6)(x - 5) = x^2 + x - 30$

Now, you need to practice factoring a trinomial into two binomials.

Example 7:   Factor $x^2 + 6x + 8$

    Step 1:   When the trinomial is in descending order as in the example above, you need to find a pair of numbers whose sum equals the number in the second term, while their product equals the third term. In the above example, find the pair of numbers that has a sum of 6 and a product of 8.

$$\underline{\quad\quad} + \underline{\quad\quad} = 6 \quad\quad \text{and} \quad\quad \underline{\quad\quad} \times \underline{\quad\quad} = 8$$

The pair of numbers that satisfy both equations is 4 and 2.

    Step 2:   Use the pair of numbers in the binomials.

The factors of $x^2 + 6x + 8$ are $(x + 4)(x + 2)$

    Check: To check, use the FOIL method.
$(x + 4)(x + 2) = x^2 + 4x + 2x + 8 = x^2 + 6x + 8$

Notice, when the second term and the third term of the trinomial are both positive, both numbers in the solution are positive.

Example 8:   Factor $x^2 - x - 6$ Find the pair of numbers where:

the sum is -1 and the product is -6

$$\underline{\quad\quad} + \underline{\quad\quad} = -1 \quad\quad \text{and} \quad\quad \underline{\quad\quad} \times \underline{\quad\quad} = -6$$

The pair of numbers that satisfies both equations is 2 and -3.
The factors of $x^2 - x - 6$ are $(x + 2)(x - 3)$

Notice, if the second term and the third term are negative, one number in the solution pair is positive, and the other number is negative.

Example 9:   Factor $x^2 - 7x + 12$ Find the pair of numbers where:

the sum is -7 and the product is 12

_____ + _____ = -7 and   _____ × _____ = 12

The pair of numbers that satisfies both equations is -3 and -4
The factors of $x^2 - 7x + 12$ are $(x-3)(x-4)$.

Notice, if the second term of a trinomial is negative and the third term is positive, both numbers in the solution are negative.

Find the factors of the following trinomials.

1. $x^2 - x - 2$

2. $y^2 + y - 6$

3. $w^2 + 3w - 4$

4. $t^2 + 5t + 6$

5. $x^2 + 2x - 8$

6. $k^2 - 4k + 3$

7. $t^2 + 3t - 10$

8. $x^2 - 3x - 4$

9. $y^2 - 5y + 6$

10. $y^2 + y - 20$

11. $a^2 - a - 6$

12. $b^2 - 4b - 5$

13. $c^2 - 5c - 14$

14. $c^2 - c - 12$

15. $d^2 + d - 6$

16. $x^2 - 3x - 28$

17. $y^2 + 3y - 18$

18. $a^2 - 9a + 20$

19. $b^2 - 2b - 15$

20. $c^2 + 7c - 8$

21. $t^2 - 11t + 30$

22. $w^2 + 13w + 36$

23. $m^2 - 2m - 48$

24. $y^2 + 14y + 49$

25. $x^2 + 7x + 10$

26. $a^2 - 7a + 6$

27. $d^2 - 6d - 27$

15.6   Factoring More Trinomials

Some trinomials have a whole number in front of the first term that cannot be factored out of the trinomial. The trinomial can still be factored.

Example 10:   Factor $2x^2 + 5x - 3$

Step 1:   To get a product of $2x^2$, one factor must begin with $2x$ and the other with x.

$(2x \quad)(x \quad)$

Step 2:   Now think: What two numbers give a product of -3? The two possibilities are 3 and -1 or -3 and 1. We know they could be in any order so there are 4 possible arrangements.

$(2x + 3)(x - 1)$
$(2x - 3)(x + 1)$
$(2x + 1)(x - 3)$
$(2x - 1)(x + 3)$

Step 3:   Multiply each possible answer until you find the arrangement of the numbers that works. Multiply the outside terms and the inside terms and add them together to see which one will equal $5x$.

$(2x + 3)(x - 1) = 2x^2 + x - 3$
$(2x - 3)(x + 1) = 2x^2 - x - 3$
$(2x + 1)(x - 3) = 2x^2 - 5 - 3$
$\boxed{(2x - 1)(x + 3) = 2x^2 + 5x - 3}$ $\longleftarrow$ This arrangement works, therefore:

The factors of $2x^2 + 5x - 3$ are $(2x - 1)(x + 3)$

Alternative: You can do some of the multiplying in your head. For the above example, ask yourself the following question: What two numbers give a product of -3 and give a sum of 5 (the whole number in the second term) when one number is first multiplied by 2 (the whole number in front of the first term)? The pair of numbers, -1 and 3, have a product of -3 and a sum of 5 when the 3 is first multiplied by 2. Therefore, the 3 will go in the opposite factor of the $2x$ so that when the terms are multiplied, you get -5.

You can use this method to at least narrow down the possible pairs of numbers when you have several from which to choose.

Factor the following trinomials.

1. $3y^2 + 14y + 8$

2. $5a^2 + 24a - 5$

3. $7b^2 + 30b + 8$

4. $2c^2 - 9c + 9$

5. $2y^2 - 7y - 15$

6. $3x^2 + 4x + 1$

7. $7y^2 + 13y - 2$

8. $11a^2 + 35a + 6$

9. $5y^2 + 17y - 12$

10. $3a^2 + 4a - 7$

11. $2a^2 + 3a - 20$

12. $5b^2 - 13b - 6$

13. $3y^2 - 4y - 32$

14. $2x^2 - 17x + 36$

15. $11x^2 - 29x - 12$

16. $5c^2 + 2c - 16$

17. $7y^2 - 30y + 27$

18. $2x^2 - 3x - 20$

19. $5b^2 + 24b - 5$

20. $7d^2 + 27d + 18$

21. $3x^2 - 20x + 25$

22. $2a^2 - 7a - 4$

23. $5m^2 + 12m + 4$

24. $9y^2 - 5y - 4$

25. $2b^2 - 13b + 18$

26. $7x^2 + 31x - 20$

27. $3c^2 - 2c - 21$

15.7   Factoring Trinomials with Two Variables

Some trinomials have two variables with exponents. You can still factor these trinomials.

Example 11:   Factor $x^2 + 5xy + 6y^2$

Step 1:   Notice there is an x^2 in the first term and a y^2 in the last term. When you see two different terms that are squared, you know there has to be an x and a y in each factor:

$$(x \quad y)(x \quad y)$$

Step 2:   Now think: What are two numbers whose sum is 5 and product is 6? You see that 3 and 2 will work. Put 3 and 2 in the factors:

$$(x + 3y)(x + 2y)$$

Check: Multiply to check. $(x + 3y)(x + 2y) = x^2 + 3xy + 2xy + 6y^2 = x^2 + 5xy + 6y^2$

Factor the following trinomials.

1. $a^2 + 6ab + 8b^2$

2. $x^2 + 3xy - 4y^2$

3. $c^2 - 2cd - 15d^2$

4. $g^2 + 7gh + 10h^2$

5. $a^2 - 5ab + 6b^2$

6. $c^2 - cd - 30d^2$

7. $x^2 + 5xy - 24y^2$

8. $a^2 - 4ab + 4b^2$

9. $c^2 - 11cd + 30d^2$

10. $x^2 - 6xy + 8y^2$

11. $g^2 - gh - 42h^2$

12. $a^2 - ab - 20b^2$

13. $x^2 + 12xy + 32y^2$

14. $c^2 + 3cd - 40d^2$

15. $x^2 + 6xy - 27y^2$

16. $a^2 - 2ab - 48b^2$

17. $c^2 - 3cd - 28d^2$

18. $x^2 + xy - 6y^2$

15.8   Factoring the Difference of Two Squares

Let's give an example of a **perfect square**.

25 is a perfect square because $5 \times 5 = 25$
49 is a perfect square because $7 \times 7 = 49$

Any variable with an even exponent is a perfect square.

y^2 is a perfect square because $y \times y = y^2$
y^4 is a perfect square because $y^2 \times y^2 = y^4$

When two terms that are both perfect squares are subtracted, factoring those terms is very easy. To factor the difference of perfect squares, you use the square root of each term, a plus sign in the first factor, and a minus sign in the second factor.

Example 12:   Factor $4x^2 - 9$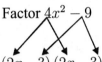

This example has two terms which are both perfect squares, and the terms are subtracted.

Step 1: $(2x \quad 3)(2x \quad 3)$

Find the square root of each term.
Use the square roots in each of the factors.

Step 2: $(2x + 3)(2x - 3)$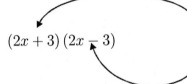

Use a plus sign in one factor and a minus sign in the other factor.

Check: Multiply to check. $(2x + 3)(2x - 3) = 4x^2 - 6x + 6x - 9 = 4x^2 - 9$

The inner and outer terms add to zero.

Example 13:   Factor $81y^4 - 1$

Step 1: $(9y^2 + 1)(9y^2 - 1)$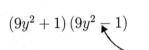

Factor like the example above.
Notice, the second factor is also the difference of two perfect squares.

Step 2: $(9y^2 + 1)(3y + 1)(3y - 1)$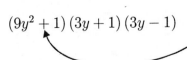

Factor the second term further.
Note: You cannot factor the sum of two perfect squares.

Check: Multiply in reverse to check your answer.
$(9y^2 + 1)(3y + 1)(3y - 1) = (9y^2 + 1)(9y^2 - 3y + 3y - 1) =$
$(9y^2 + 1)(9y^2 - 1) = 81y^4 + 9y^2 - 9y^2 - 1 = 81y^4 - 1$

Factor the following differences of perfect squares.

1. $64x^2 - 49$

2. $4y^4 - 25$

3. $9a^4 - 4$

4. $25c^4 - 9$

5. $64y^2 - 9$

6. $x^4 - 16$

7. $49x^2 - 4$

8. $4d^2 - 25$

9. $9a^2 - 16$

10. $100y^4 - 49$

11. $c^4 - 36$

12. $36x^2 - 25$

13. $25x^2 - 4$

14. $9x^4 - 64$

15. $49x^2 - 100$

16. $16x^2 - 81$

17. $9y^4 - 1$

18. $49c^2 - 25$

19. $25d^2 - 64$

20. $36a^4 - 49$

21. $16x^4 - 16$

22. $b^2 - 25$

23. $c^4 - 144$

24. $9y^2 - 4$

25. $81x^4 - 16$

26. $4b^2 - 36$

27. $9w^2 - 9$

28. $64a^2 - 25$

29. $49y^2 - 121$

30. $x^6 - 9$

15.9   Factoring Using Special Products

In Algebra, we use special products to factor expressions. There are only 6 formulas for special products and they are:

1. $x^2 + 2xy + y^2 = (x + y)^2$ Square of a sum
2. $x^2 - 2xy + y^2 = (x - y)^2$ Square of a difference
3. $x^2 - y^2 = (x + y)(x - y)$ Product of a sum and difference
4. $x^2 + (a + b)x + ab = (x + a)(x + b)$ Product of a sum/difference with two different numbers
5. $x^3 + 3x^2y + 3xy^2 + y^3 = (x + y)^3$ Cube of a sum
6. $x^3 - 3x^2y + 3xy^2 - y^3 = (x - y)^3$ Cube of a difference

Using the formulas for special products, we can quickly factor these problems.

Example 14: $x^2 + 6x + 9 = (x + 3)^2$

Example 15: $x^2 - 6x + 9 = (x - 3)^2$

Example 16: $x^2 - 9 = (x + 3)(x - 3)$

Example 17: $x^2 + 5x + 5 = (x + 3)(x + 2)$

Example 18: $x^3 + 9x^2 + 27x + 27 = (x + 3)^3$

Example 19: $x^3 - 9x^2 + 27x - 27 = (x - 3)^3$

Use formulas for special products to expand the binomials.

1. $(x + 2)^2$	4. $(x - 9)^2$	7. $(x + 4)(x + 5)$	10. $(x + 4)^3$
2. $(x + 7)^2$	5. $(x + 7)(x - 7)$	8. $(x + 7)(x + 2)$	11. $(x - 8)^3$
3. $(x - 6)^2$	6. $(x + 5)(x - 5)$	9. $(x + 6)^3$	12. $(x - 1)^3$

Use what you know about special products to factor the expressions.

13. $x^2 + 6x + 9$	19. $x^2 + 15x + 50$
14. $x^2 + 20x + 100$	20. $x^2 + 11x + 28$
15. $x^2 - 14x + 49$	21. $x^3 + 6x^2 + 12x + 8$
16. $x^2 - 18x + 81$	22. $x^3 + 12x^2 + 48x + 64$
17. $x^2 - 1$	23. $x^3 - 9x^2 + 27x - 27$
18. $x^2 - 121$	24. $x^3 - 15x^2 + 75x - 125$

15.10   Simplifying Rational Expressions

We will use what we learned so far in this chapter to factor the terms in the numerator and the denominator when possible, then simplify the rational expression.

Example 20:   Simplify $\dfrac{c^2 - 25}{c^2 + 5c}$

Step 1:   The numerator is the difference of two perfect squares, so it can be easily factored as in the previous section. Use the square root of each of the terms in the parentheses, with a plus sign in one and a minus sign in the other.
$c^2 - 25 = (c - 5)(c + 5)$

Step 2:   Find the greatest common factor in the denominator and factor it out. In this case, it is the variable c.
$c^2 + 5c = c(c + 5)$

Step 3:   Simplify $\dfrac{c^2 - 25}{c^2 + 5c} = \dfrac{(c - 5)(c + 5)}{c(c + 5)} = \dfrac{c - 5}{c}$

Simplify the rational expressions. Check for perfect squares and common factors.

1. $\dfrac{25x^2 - 4}{5x^2 - 2x}$

5. $\dfrac{9a^2 - 16}{3a^2 - 4a}$

9. $\dfrac{4y^2 - 36}{2y^2 + 6y}$

2. $\dfrac{64c^2 - 25}{8c^2 + 5c}$

6. $\dfrac{16x^2 - 81}{4x^2 - 9x}$

10. $\dfrac{81y^4 - 16}{9y^2 + 4}$

3. $\dfrac{36a^2 - 49}{6a^2 - 7a}$

7. $\dfrac{49x^2 - 100}{7x^2 + 10x}$

11. $\dfrac{25x^4 - 225}{5x^2 + 15}$

4. $\dfrac{x^2 - 9}{x^2 + 3x}$

8. $\dfrac{x^4 - 16}{x^2 + 2x}$

12. $\dfrac{3y^3 + 9}{y^9 - 9}$

15.11   Adding Rational Expressions

Rational expressions are fractions that can have variables in the numerator and/or the denominator. Adding rational expressions is similar to adding fractions.

Example 21:   Add: $\dfrac{x-3}{x-2} + \dfrac{5}{x+3}$

Step 1:   Just like adding fractions, we must find a common denominator by the finding the least common multiple of the denominators. The least common multiple of $x-2$ and $x+3$ is $(x-2)(x+3)$.

Step 2:   Set up the algebra problem like a fraction problem and find the numerators.

$$\frac{x-3}{x-2} = \frac{(x-3)(x+3)}{(x-2)(x+3)}$$

$$+ \quad \frac{5}{x+3} = \frac{5x-10}{(x+3)(x-2)}$$

Step 3:   Add: $\dfrac{(x-3)(x+3)+5x-10}{(x+3)(x-2)}$

Step 4:   Simplify: $\dfrac{x^2-9+5x-10}{(x+3)(x-2)} = \dfrac{x^2+5x-19}{(x+3)(x-2)}$

Add the following rational expressions.

1. $\dfrac{x^2-9x-22}{x^2-8x-33} + \dfrac{x^2+12x+35}{x^2+10x+21}$

2. $\dfrac{1}{x^3} + \dfrac{x-9}{x^2}$

3. $\dfrac{1}{2x} + \dfrac{x^2-3}{2x}$

4. $\dfrac{y^3}{x} + \dfrac{x^3}{y}$

5. $\dfrac{(x-2)(x+13)}{x^2-4} + \dfrac{(x-3)(x+1)}{x^2-x-6}$

6. $\dfrac{a}{b^2} + \dfrac{b^3}{a-b}$

7. $\dfrac{5}{y^3} + \dfrac{7}{y^3}$

8. $\dfrac{12b}{a^2} + \dfrac{31c}{a^2}$

9. $\dfrac{y}{x^2} + \dfrac{3}{y^2} + \dfrac{x}{x^2y^2}$

10. $\dfrac{r}{q^{12}} + \dfrac{s}{r^{15}} + \dfrac{q}{s^5}$

11. $\dfrac{x}{x-y} + \dfrac{y}{x-y}$

12. $\dfrac{9}{ba} + \dfrac{2}{b}$

15.12   Subtracting Rational Expressions

Subtracting rational expressions is similar to subtracting fractions.

Example 22:   Subtract: $\dfrac{x^2}{5} - \dfrac{9}{x+3}$

   Step 1:   Just like adding we must find a common denominator. The common denominator of the rational expressions is $5\,(x+3)$.

   Step 2:   Set up the algebra problem like a fraction problem and find the numerators.

$$\dfrac{x^2}{5} = \dfrac{x^2\,(x+3)}{5\,(x+3)}$$

$$-\ \dfrac{9}{x+3} = \dfrac{45}{5\,(x+3)}$$

   Step 3:   Subtract: $\dfrac{x^2\,(x+3) - 45}{5\,(x+3)}$

   Step 4:   Simplify: $\dfrac{x^2\,(x+3) - 45}{5\,(x+3)} = \dfrac{x^3 + 3x^2 - 45}{5\,(x+3)}$

Subtract the following rational expressions.

1. $\dfrac{x}{3y} - \dfrac{2}{3y}$

2. $\dfrac{7}{5c} - \dfrac{2}{c}$

3. $\dfrac{x}{y^2} - \dfrac{x^2}{y^2}$

4. $\dfrac{x+1}{y+1} - \dfrac{y+1}{x+1}$

5. $\dfrac{1}{x} - \dfrac{2x^2}{y}$

6. $\dfrac{2a^2}{bc} - \dfrac{a^5}{bc}$

7. $\dfrac{d-e}{a+b} - \dfrac{5}{a+b}$

8. $\dfrac{x}{5y^2} - \dfrac{y}{5x^2}$

9. $\dfrac{c^3}{3b} - \dfrac{b^2}{4a}$

10. $\dfrac{2x+5}{y} - \dfrac{5x^3 - x^2}{y}$

11. $\dfrac{2d}{3c^5} - \dfrac{a^5}{2b^2}$

12. $\dfrac{5x}{y} - \dfrac{2x}{y}$

15.13 Multiplying Rational Expressions

Multiplying rational expressions is similar to multiplying fractions because a rational expression is a fraction.

Example 23: Multiply: $\dfrac{x^4 - x^3}{y^3} \times \dfrac{y^4}{x^2}$

Step 1: When you multiply a rational expression, you must multiply the numerators together and multiply the denominators together.
$$\frac{x^4 - x^3}{y^3} \times \frac{y^4}{x^2} = \frac{(x^4 - x^3) \times y^4}{y^3 \times x^2} = \frac{x^4 y^4 - x^3 y^4}{x^2 y^3}$$

Step 2: Simplify the resulting rational expression. You can factor $x^3 y^4$ out of the numerator.
$$\frac{x^4 y^4 - x^3 y^4}{x^2 y^3} = \frac{x^3 y^4 (x - 1)}{x^2 y^3}$$

Step 3: You can also cancel $x^2 y^3$ because it is in the numerator and denominator of the expression.
$$\frac{x^3 y^4 (x - 1)}{x^2 y^3} = \frac{xy(x - 1)}{1} = xy(x - 1)$$

Therefore, $\dfrac{x^4 - x^3}{y^3} \times \dfrac{y^4}{x^2} = xy(x - 1)$.

Multiply the following rational expressions.

1. $\dfrac{-b}{2a} \times \dfrac{-a}{3b}$

2. $\dfrac{4x}{y} \times \dfrac{1}{2y}$

3. $\dfrac{5a}{3b} \times \dfrac{4c}{3b}$

4. $\dfrac{a + b}{a - b} \times \dfrac{c + b}{c - b}$

5. $\dfrac{c}{a^2} \times \dfrac{c - a}{b - c}$

6. $\dfrac{y^5}{x^3} \times \dfrac{y^2 + 2y + 1}{x^2 - y}$

7. $\dfrac{x^2 - 2x - 3}{x^2 - 5x - 14} \times \dfrac{x^2 - 2x - 35}{x^2 + 6x - 27}$

8. $\dfrac{9}{x} \times \dfrac{x^5}{y}$

9. $\dfrac{a^2 - b}{a - b^2} \times \dfrac{7c}{b}$

10. $\dfrac{5}{x^2 y} \times \dfrac{4}{xy^2}$

11. $\dfrac{5x^3}{2} \times \dfrac{2x^3}{5}$

12. $\dfrac{x + 7}{x - 1} \times \dfrac{x - 3}{x + 5}$

13. $\dfrac{3c}{a^2} \times \dfrac{2ba}{c}$

14. $\dfrac{a^5 - a^2}{b^3} \times \dfrac{c^4}{a^2}$

15. $\dfrac{b - 17}{c^2 + 2} \times \dfrac{a^3}{b^3} \times \dfrac{c^3}{b - 17}$

16. $\dfrac{hk}{47} \times \dfrac{16}{m}$

15.14   Dividing Rational Expressions

Dividing rational expressions is similar to dividing fractions because a rational expression is a fraction.

Example 24:   Divide: $\dfrac{2a^3}{b} \div \dfrac{5}{b^2}$

Step 1:   Just like dividing fractions, you must flip the second expression to get its reciprocal, then multiply.

$$\frac{2a^3}{b} \div \frac{5}{b^2} = \frac{2a^3}{b} \times \frac{b^2}{5}$$

Step 2:   Multiply the two rational expressions together just like you did in the previous section.

$$\frac{2a^3}{b} \times \frac{b^2}{5} = \frac{2a^3 \times b^2}{b \times 5} = \frac{2a^3 b^2}{5b}$$

Step 3:   Simplify: $\dfrac{2a^3 b^2}{5b} = \dfrac{2a^3 b}{5}$

Therefore, $\dfrac{2a^3}{b} \div \dfrac{5}{b^2} = \dfrac{2a^3 b}{5}$.

Divide the following rational expressions.

1. $\dfrac{x}{y} \div \dfrac{y}{x}$

2. $\dfrac{2x}{y^3} \div \dfrac{x}{y}$

3. $\dfrac{x-1}{y+2} \div \dfrac{y-5}{x+7}$

4. $\dfrac{a-2}{c-1} \div \dfrac{a-2}{c+3}$

5. $\dfrac{x^2 + 2x + 1}{y^2 + 8y + 15} \div \dfrac{x^2 + 10x + 9}{y^2 + 10y + 21}$

6. $\dfrac{a}{c} \div \dfrac{c}{b}$

7. $\dfrac{c}{a} \div \dfrac{c}{b}$

8. $\dfrac{2x+3}{x+3} \div \dfrac{x+3}{2x+3} \div \dfrac{x}{y}$

9. $\dfrac{ac}{b^2} \div \dfrac{c^2}{ab}$

10. $\dfrac{b^2}{a} \div \dfrac{a^2}{c}$

11. $\dfrac{c^2}{3b} \div \dfrac{16}{a}$

12. $\dfrac{20cd}{b} \div \dfrac{2ac}{d}$

13. $\dfrac{x+4}{x+2} \div \dfrac{x-1}{x-3}$

14. $\dfrac{y}{3x^2} \div \dfrac{2y}{x^3}$

15. $\dfrac{y^3 - 1}{x^2 + 1} \div \dfrac{x-7}{y+2}$

16. $\dfrac{200a}{7b} \div \dfrac{10a}{b}$

Chapter 15 Review

Factor the following polynomials completely.

1. $8x - 18$

2. $6x^2 - 18x$

3. $16b^3 + 8b$

4. $15a^3 + 40$

5. $20y^6 - 12y^4$

6. $5a - 15a^2$

7. $4y^2 - 36$

8. $25a^4 - 49b^2$

9. $3ax + 3ay + 4x + 4y$

10. $ax - 2x + ay - 2y$

11. $2bx + 2x - 2by - 2y$

12. $2b^2 - 2b - 12$

13. $7x^3 + 14x - 3x^2 - 6$

14. $3a^3 + 4a^2 + 9a + 12$

15. $27y^2 + 42y - 5$

16. $12b^2 + 25b - 7$

17. $c^2 + cd - 20d^2$

18. $x^2 - 4xy - 21y^2$

19. $6y^2 + 30y + 36$

20. $2b^2 + 6b - 20$

21. $16b^4 - 81d^4$

22. $9w^2 - 54w - 63$

23. $m^2p^2 - 5mp + 2m^2p - 10m$

24. $12x^2 + 27x$

25. $2xy - 36 + 8y - 9x$

26. $2a^4 - 32$

27. $21c^2 + 41c + 10$

28. $x^2 - y + xy - x$

29. $2b^3 - 24 + 16b - 3b^2$

30. $5 - 2a - 25a^2 + 10a^3$

Simplify the following rational expressions by performing the appropriate operation.

31. $\dfrac{b}{a} + \dfrac{b}{a}$

32. $\dfrac{x}{y} + \dfrac{y}{x}$

33. $\dfrac{x}{x+y} - \dfrac{y}{x+y}$

34. $\dfrac{3x+5}{x+1} - \dfrac{3x-1}{x-1}$

35. $\dfrac{-b}{ac} \times \dfrac{3b}{a-2} \times \dfrac{c}{b+1}$

36. $\dfrac{x}{y^3} \times \dfrac{3y^2}{x^2}$

37. $\dfrac{-b}{5ac} \div \dfrac{d}{b^2}$

38. $\dfrac{5}{a+3} \div \dfrac{7a+21}{b^2+b}$

Use your knowledge of special products to factor the following expressions.

39. $x^2 + 12x + 36$

40. $x^2 - 8x + 16$

41. $x^2 - 4$

42. $x^2 + 8x + 12$

43. $x^3 + 6x^2 + 12x + 8$

44. $x^3 - 12x^2 + 48x - 64$

45. $x^3 + 18x^2 + 108x + 216$

46. $x^3 + 12x^2 + 48x + 64$

47. $x^3 + 21x^2 + 147x + 343$

Chapter 15 Test

1. What is the greatest common factor of $4x^3$ and $8x^2$?

   (A) $4x^2$
   (B) $4x$
   (C) x^2
   (D) $8x$

2. Factor: $8x^4 - 7x^2 + 4x$

   (A) $4x\left(2x^3 - 7x + 4\right)$
   (B) $x\left(8x^4 - 7x^2 + 4x\right)$
   (C) $x\left(8x^3 - 7x + 4\right)$
   (D) $4x\left(2x^3 - 7x + 1\right)$

3. Factor by grouping:

   $xy + 2x + 3y + 6$

   (A) $(x + 2)(y + 3)$
   (B) $(x + 3)(y + 2)$
   (C) $x(y + 2) + 3(y + 2)$
   (D) This problem cannot be factored.

4. Factor: $x^2 + 6x + 8$

   (A) $(x + 2)(x + 4)$
   (B) $(x + 1)(x + 8)$
   (C) $(x - 2)(x - 4)$
   (D) $(x - 1)(x - 8)$

5. Divide: $\dfrac{5}{2x} \div \dfrac{5x}{y}$

   (A) $\dfrac{y}{2x}$

   (B) $\dfrac{25}{2x^2 y}$

   (C) $\dfrac{y}{2x^2}$

   (D) $\dfrac{25x}{2y}$

6. Simplify the following rational expression:

   $$\frac{36x^4 - 16}{6x^3 + 4x}$$

   (A) $\dfrac{\left(6x^2 - 4\right)\left(6x^2 + 4\right)}{x}$

   (B) $\dfrac{6x^2 - 4}{x}$

   (C) $6x^2 - 4$

   (D) $\dfrac{(3x - 2)(3x + 2)}{x}$

7. Simplify the following rational expression:

   $$\frac{c^2 + 10c + 24}{c^3 + 4c^2}$$

   (A) $\dfrac{c + 6}{c}$

   (B) $\dfrac{c + 4}{c^2}$

   (C) $\dfrac{c + 6}{c^2}$

   (D) $\dfrac{(c + 6)(c + 4)}{c^2(c + 4)}$

8. Add: $\dfrac{3x^3}{y} + \dfrac{3y^3}{x}$

   (A) $\dfrac{9x^3 y^3}{xy}$

   (B) $\dfrac{3x^4 + 3y^4}{xy}$

   (C) $\dfrac{3x^3 + 3y^3}{x + y}$

   (D) $\dfrac{3x^4 + 3y^4}{x + y}$

9. Add: $\dfrac{11x - y}{y} + \dfrac{2}{xy^2}$

   (A) $\dfrac{11x^2y - xy^2 + 2}{xy^3}$

   (B) $\dfrac{11x^2y - xy^2 - 2}{xy^2}$

   (C) $\dfrac{22x - 2y}{y + xy^2}$

   (D) $\dfrac{11x^2y - xy^2 + 2}{xy^2}$

10. Subtract: $\dfrac{x + 2}{x^2 - 4} - \dfrac{x}{x - 2}$

   (A) $\dfrac{1 - x}{x - 2}$

   (B) $\dfrac{x}{x^2 - 4}$

   (C) $\dfrac{2x - 2}{x^2 - 4}$

   (D) $\dfrac{x + 2}{x^2 - x - 6}$

11. Multiply: $\dfrac{x + 3}{2x} \times \dfrac{x\,(y - 1)}{y}$

   (A) $\dfrac{xy - x + 3y - 3}{2x}$

   (B) $\dfrac{xy - x + 3y - 3}{2xy}$

   (C) $\dfrac{xy - x + 3y - 3}{xy}$

   (D) $\dfrac{xy - x + 3y - 3}{2y}$

12. Factor: $2x^2 - 2x - 84$

   (A) $(2x + 7)\,(x - 12)$
   (B) $(2x - 12)\,(x + 7)$
   (C) $(2x - 7)\,(x + 12)$
   (D) $(2x + 12)\,(x - 7)$

13. Subtract: $\dfrac{1}{7x} - \dfrac{13y}{x}$

   (A) $\dfrac{14y}{8x}$

   (B) $-\dfrac{13y}{7x}$

   (C) $\dfrac{1 - 13y}{6x}$

   (D) $\dfrac{1 - 91y}{7x}$

14. Multiply: $\dfrac{12y}{x} \times \dfrac{y^3}{x}$

   (A) $\dfrac{12y^4}{x^2}$

   (B) $\dfrac{12y}{x^3}$

   (C) $\dfrac{36y^4}{x^2}$

   (D) $\dfrac{12y^4}{x}$

15. Divide: $\dfrac{x - 2}{y} \div \dfrac{7}{x + 3}$

   (A) $\dfrac{7x - 14}{y + x + 3}$

   (B) $\dfrac{x^2 + x - 6}{7y}$

   (C) $\dfrac{7x - 14}{xy + 3y}$

   (D) $\dfrac{x^2 + x - 6}{y}$

16. Factor: $4x^2 - 64$

   (A) $(x - 8)\,(x + 8)$
   (B) $(4x - 8)\,(4x + 8)$
   (C) $(2x - 16)\,(2x + 16)$
   (D) $(2x - 8)\,(2x + 8)$

17. Factor: $x^2 + 22x + 121$

(A) $(x + 12)^2$
(B) $(x - 11)^2$
(C) $(x + 7)^2$
(D) $(x + 11)^2$

18. Factor: $x^2 - 24x + 144$

(A) $(x + 12)^2$
(B) $(x - 12)^2$
(C) $(x + 12)(x - 12)$
(D) $(x + 12)^3$

19. Factor: $x^2 - 36$

(A) $(x + 7)(x + 4)$
(B) $(x + 4)(x - 9)$
(C) $(x + 6)(x - 6)$
(D) $(x + 6)(x + 6)$

20. Factor: $x^3 + 18x^2 + 108x + 216$

(A) $(x + 6)^3$
(B) $(x + 4)^3$
(C) $(x + 8)^3$
(D) $(x - 6)^3$

21. Factor: $x^3 - 27x^2 + 243x - 729$

(A) $(x - 11)^3$
(B) $(x - 9)^3$
(C) $(x - 7)^3$
(D) $(x - 8)^3$

22. What are the factors of $x^2 + 10x + 25$?

(A) $(x + 5)^2$
(B) $(x - 5)^2$
(C) $(x - 5)(x + 5)$
(D) $(x + 5)(x + 2)$

23. What are the factors of $x^2 + 11x + 30$?

(A) $(x + 6)(x + 5)$
(B) $(x + 10)(x + 3)$
(C) $(x + 6)^2$
(D) $(x + 15)(x + 2)$

24. What are the factors of $x^3 - 3x^2 + 3x - 1$?

(A) $(x + 1)^3$
(B) $(x + 3)^3$
(C) $(x - 3)^3$
(D) $(x - 1)^3$

25. What are the factors of $x^3 + 21x^2 + 147x + 343$?

(A) $(x + 7)^3$
(B) $(x - 7)^3$
(C) $(x - 3)^3$
(D) $(x + 9)^3$

Chapter 16
Solving Quadratic Equations and Inequalities

This chapter covers the following Georgia Performance Standards:

MA1A	Algebra	MA1A3c
		MA1A4a
		MA1A4b
		MA1A4c
		MA1A4d

You can factor polynomials such as $y^2 - 4y - 5$ into two factors:

$$y^2 - 4y - 5 = (y + 1)(y - 5)$$

In this chapter, we learn that any equation that can be put in the form $ax^2 + bx + c = 0$ is a quadratic equation if a, b, and c are real numbers and $a \neq 0$. $ax^2 + bx + c = 0$ is the standard form of a quadratic equation. To solve these equations, follow the steps below.

Example 1: Solve $y^2 - 4y - 5 = 0$

Step 1: Factor the left side of the equation.

$$\begin{aligned} y^2 - 4y - 5 &= 0 \\ (y + 1)(y - 5) &= 0 \end{aligned}$$

Step 2: If the product of these two factors equals zero, then the two factors individually must be equal to zero. Therefore, to solve, we set each factor equal to zero.

$$\begin{array}{rl} (y + 1) &= 0 \\ -1 \quad -1 & \\ \hline y &= -1 \end{array} \qquad \begin{array}{rl} (y - 5) &= 0 \\ +5 \quad +5 & \\ \hline y &= 5 \end{array}$$

The equation has two solutions: $y = -1$ and $y = 5$

Check: To check, substitute each solution into the original equation.

When $y = -1$, the equation becomes:
$$(-1)^2 - (4)(-1) - 5 = 0$$
$$1 + 4 - 5 = 0$$
$$0 = 0$$

When $y = 5$, the equation becomes:
$$5^2 - (4)(5) - 5 = 0$$
$$25 - 20 - 5 = 0$$
$$0 = 0$$

Both solutions produce true statements.
The solution set for the equation is $\{-1, -5\}$.

Solve each of the following quadratic equations by factoring and setting each factor equal to zero. Check by substituting answers back in the original equation.

1. $x^2 + x - 6 = 0$

2. $y^2 - 2y - 8 = 0$

3. $a^2 + 2a - 15 = 0$

4. $y^2 - 5y + 4 = 0$

5. $b^2 - 9b + 14 = 0$

6. $x^2 - 3x - 4 = 0$

7. $y^2 + y - 20 = 0$

8. $d^2 + 6d + 8 = 0$

9. $y^2 - 7y + 12 = 0$

10. $x^2 - 3x - 28 = 0$

11. $a^2 - 5a + 6 = 0$

12. $b^2 + 3b - 10 = 0$

13. $a^2 + 7a - 8 = 0$

14. $x^2 + 3x + 2 = 0$

15. $x^2 - x - 42 = 0$

16. $a^2 + a - 6 = 0$

17. $b^2 + 7b + 12 = 0$

18. $y^2 + 2y - 15 = 0$

19. $a^2 - 3a - 10 = 0$

20. $d^2 + 10d + 16 = 0$

21. $x^2 - 4x - 12 = 0$

Quadratic equations that have a whole number and a variable in the first term are solved the same way as the previous page. Factor the trinomial, and set each factor equal to zero to find the solution set.

Example 2: Solve $2x^2 + 3x - 2 = 0$
$(2x - 1)(x + 2) = 0$
Set each factor equal to zero and solve:

$$
\begin{array}{ll}
2x - 1 = 0 & \\
\underline{+1 \quad +1} & \\
\dfrac{2x}{2} = \dfrac{1}{2} & \\
x = \dfrac{1}{2} &
\end{array}
\qquad
\begin{array}{ll}
x + 2 = 0 \\
\underline{-2 \quad -2} \\
x = -2
\end{array}
$$

The solution set is $\left\{ \dfrac{1}{2}, -2 \right\}$.

Solve the following quadratic equations.

22. $3y^2 + 4y - 32 = 0$

23. $5c^2 - 2c - 16 = 0$

24. $7d^2 + 18d + 8 = 0$

25. $3a^2 - 10a - 8 = 0$

26. $11x^2 - 31x - 6 = 0$

27. $5b^2 + 17b + 6 = 0$

28. $3x^2 - 11x - 20 = 0$

29. $5a^2 + 47a - 30 = 0$

30. $2c^2 - 5c - 25 = 0$

31. $2y^2 + 11y - 21 = 0$

32. $5a^2 + 23a - 42 = 0$

33. $3d^2 + 11d - 20 = 0$

34. $3x^2 - 10x + 8 = 0$

35. $7b^2 + 23b - 20 = 0$

36. $9a^2 - 58a + 24 = 0$

37. $4c^2 - 25c - 21 = 0$

38. $8d^2 + 53d + 30 = 0$

39. $4y^2 + 37y - 30 = 0$

40. $8a^2 + 37a - 15 = 0$

41. $3x^2 - 41x + 26 = 0$

42. $8b^2 + 2b - 3 = 0$

16.1   Using the Quadratic Formula

You may be asked to use the quadratic formula to solve an algebra problem known as a **quadratic equation**. The equation should be in the form $ax^2 + bx + c = 0$.

Example 3:   Using the quadratic formula, find x in the following equation: $x^2 - 8x = -7$.

Step 1:   Make sure the equation is set equal to 0.

$$x^2 - 8x + 7 = -7 + 7$$
$$x^2 - 8x + 7 = 0$$

The quadratic formula, $\dfrac{-b \pm \sqrt{b^2 - 4ac}}{2a}$, will be given to you on your formula sheet with your test.

Step 2:   In the formula, a is the number x^2 is multiplied by, b is the number x is multiplied by and c is the last term of the equation. For the equation in the example, $x^2 - 8x + 7$, $a = 1$, $b = -8$, and $c = 7$. When we look at the formula we notice a $\pm$ sign. This means that there will be two solutions to the equation, one when we use the plus sign and one when we use the minus sign. Substituting the numbers from the problem into the formula, we have:

$$\frac{8 + \sqrt{8^2 - (4)(1)(7)}}{2(1)} = 7 \qquad \text{or} \qquad \frac{8 - \sqrt{8^2 - (4)(1)(7)}}{2(1)} = 1$$

The solutions are $\{7, 1\}$.

For each of the following equations, use the quadratic formula to find two solutions.

1. $x^2 + x - 6 = 0$

2. $y^2 - 2y - 8 = 0$

3. $a^2 + 2a - 15 = 0$

4. $y^2 - 5y + 4 = 0$

5. $b^2 - 9b + 14 = 0$

6. $x^2 - 3x - 4 = 0$

7. $y^2 + y - 20 = 0$

8. $d^2 + 6d + 8 = 0$

9. $y^2 - 7y + 12 = 0$

10. $x^2 - 3x - 28 = 0$

11. $a^2 - 5a + 6 = 0$

12. $b^2 + 3b - 10 = 0$

13. $a^2 + 7a - 8 = 0$

14. $c^2 + 3c + 2 = 0$

15. $x^2 - x - 42 = 0$

16. $a^2 + 5a - 6 = 0$

17. $b^2 + 7b + 12 = 0$

18. $y^2 + y - 12 = 0$

19. $a^2 - 3a - 10 = 0$

20. $d^2 + 10d + 16 = 0$

21. $x^2 - 4x - 12 = 0$

16.2   Solving Quadratic Equations with Complex Roots

Some quadratic equations do not have real number solutions. In this section, we will find complex solutions to each equation.

Example 4:   Find the roots of the quadratic equation $7x^2 - 8x + 3 = 0$.

Step 1:   The quadratic equation cannot be factored, so the quadratic formula must be used.

$$x = \frac{-b \pm \sqrt{b^2 - 4ac}}{2a} = \frac{-(-8) \pm \sqrt{(-8)^2 - 4\,(7)\,(3)}}{2\,(7)} = \frac{8 \pm \sqrt{-20}}{14}$$

Step 2:   As stated earlier, i is defined as $\sqrt{-1}$, so if $\dfrac{8 \pm \sqrt{-20}}{14}$ is rewritten as

$\dfrac{8 \pm \sqrt{20 \times (-1)}}{14}$, and it can be simplified to $\dfrac{8 \pm i\sqrt{20}}{14}$.

This can be simplified as follows:

$$\frac{8 \pm i\sqrt{20}}{14} = \frac{8 \pm i\sqrt{4 \times 5}}{14} = \frac{8 \pm 2i\sqrt{5}}{14} = \frac{4 \pm i\sqrt{5}}{7} = \frac{4}{7} \pm \frac{\sqrt{5}}{7}i.$$

Therefore, the roots of the equation are $x = \dfrac{4}{7} + \dfrac{\sqrt{5}}{7}i$ or $\dfrac{4}{7} - \dfrac{\sqrt{5}}{7}i$.

Example 5:   Find the roots of the quadratic equation $3x^2 - x + 5 = 0$.

Step 1:   Plug the values for this equation into the quadratic formula: $a = 3$, $b = -1$, $c = 5$.

Step 2:   Solve for x.

$$x = \frac{-(-1) \pm \sqrt{(-1)2 - 4(3)(5)}}{2(3)} = \frac{1 \pm \sqrt{1 - 60}}{6} = \frac{1}{6} \pm \frac{\sqrt{-59}}{6}$$

Step 3:   Apply the definition of i. $x = \dfrac{1}{6} \pm \dfrac{\sqrt{-59}}{6} = \dfrac{1}{6} \pm \dfrac{\sqrt{59}}{6}i.$

Find the roots of each of the following quadratic equations.

1. $9x^2 - 6x + 3 = 0$

2. $\frac{3}{4}x^2 - x + 12 = 0$

3. $-11x^2 + 10x - 3 = 0$

4. $14x^2 - 3x + \frac{1}{2} = 0$

5. $-x^2 - 3x - 16 = 0$

6. $8x^2 + 9x + 10 = 0$

7. $x(19x - 2) = -1$

8. $x(1 - 5x) = 5$

9. $-10x^2 + \frac{x}{2} - 7 = 0$

10. $8x^2 + 3x + 10 = 0$

11. $-6x^2 - 5x - 11 = 0$

12. $13x^2 + 12x + 7 = 0$

13. $4x^2 - 10x + 11 = 0$

14. $-5x^2 - x - 3 = 0$

15. $18x^2 - 5x + 1 = 0$

16.3   Solving the Difference of Two Squares

To solve the difference of two squares, first factor. Then set each factor equal to zero.

Example 6: $25x^2 - 36 = 0$

Step 1:   Factor the left hand side of the equation.

$25x^2 - 36 = 0$
$(5x + 6)(5x - 6) = 0$

Step 2:   Set each factor equal to zero and solve.

$$\begin{array}{ll}
\begin{aligned}
5x + 6 &= 0 \\
-6 \quad &-6 \\
\hline
\frac{5x}{5} &= \frac{6}{5} \\
x &= -\frac{6}{5}
\end{aligned}
&
\begin{aligned}
5x - 6 &= 0 \\
+6 \quad &+6 \\
\hline
\frac{5x}{5} &= \frac{6}{5} \\
x &= \frac{6}{5}
\end{aligned}
\end{array}$$

Check: Substitute each solution in the equation to check.

for $x = -\dfrac{6}{5}$:

$25x^2 - 36 = 0$

$25\left(-\dfrac{6}{5}\right)\left(-\dfrac{6}{5}\right) - 36 = 0 \longleftarrow$ Substitute $-\dfrac{6}{5}$ for x.

$25\left(\dfrac{36}{25}\right) - 36 = 0 \longleftarrow$ Cancel the 25's.

$36 - 36 = 0 \longleftarrow$ A true statement. $x = -\dfrac{6}{5}$ is a solution.

for $x = \dfrac{6}{5}$:

$25x^2 - 36 = 0$

$25\left(\dfrac{6}{5}\right)\left(\dfrac{6}{5}\right) - 36 = 0 \longleftarrow$ Substitute $\dfrac{6}{5}$ for x.

$25\left(\dfrac{36}{25}\right) - 36 = 0 \longleftarrow$ Cancel the 25's.

$36 - 36 = 0 \longleftarrow$ A true statement. $x = \dfrac{6}{5}$ is a solution.

The solution set is $\left\{\dfrac{-6}{5}, \dfrac{6}{5}\right\}$.

Find the solution sets for the following.

1. $25a^2 - 16 = 0$

2. $c^2 - 36 = 0$

3. $9x^2 - 64 = 0$

4. $100y^2 - 49 - 0$

5. $4b^2 - 81 = 0$

6. $d^2 - 25 = 0$

7. $9x^2 - 1 = 0$

8. $16a^2 - 9 = 0$

9. $36y^2 - 1 = 0$

10. $36y^2 - 25 = 0$

11. $d^2 - 16 = 0$

12. $64b^2 - 9 = 0$

13. $81a^2 - 4 = 0$

14. $64y^2 - 25 = 0$

15. $4c^2 - 49 = 0$

16. $x^2 - 81 = 0$

17. $49b^2 - 9 = 0$

18. $a^2 - 64 = 0$

19. $x^2 - 1 = 0$

20. $4y^2 - 9 = 0$

21. $t^2 - 100 = 0$

22. $16k^2 - 81 = 0$

23. $a^2 - 4 = 0$

24. $36b^2 - 16 = 0$

16.4 Solving Perfect Squares

When the square root of a constant, variable, or polynomial results in a constant, variable, or polynomial without irrational numbers, the expression is a **perfect square**. Some examples are 49, x^2, and $(x-2)^2$.

Example 7: Solve the perfect square for x. $(x-5)^2 = 0$

Step 1: Take the square root of both sides.
$$\sqrt{(x-5)^2} = \sqrt{0}$$
$$(x-5) = 0$$

Step 2: Solve the equation.
$$(x-5) = 0$$
$$x-5+5 = 0+5$$
$$x = 5$$

Example 8: Solve the perfect square for x. $(x-5)^2 = 64$

Step 1: Take the square root of both sides.
$$\sqrt{(x-5)^2} = \sqrt{64}$$
$$(x-5) = \pm 8$$
$$(x-5) = 8 \text{ and } (x-5) = -8$$

Step 2: Solve the two equations.
$$(x-5) = 8 \qquad \text{and} \quad (x-5) = -8$$
$$x-5+5 = 8+5 \quad \text{and} \quad x-5+5 = -8+5$$
$$x = 13 \qquad\qquad \text{and} \quad x = -3$$

Solve the perfect square for x.

1. $(x-2)^2 = 0$

2. $(x+1)^2 = 0$

3. $(x+11)^2 = 0$

4. $(x-4)^2 = 0$

5. $(x-1)^2 = 0$

6. $(x+8)^2 = 0$

7. $(x+3)^2 = 4$

8. $(x-5)^2 = 16$

9. $(x-10)^2 = 100$

10. $(x+9)^2 = 9$

11. $(x-4.5)^2 = 25$

12. $(x+7)^2 = 36$

13. $(x+2)^2 = 49$

14. $(x-1)^2 = 4$

15. $(x+8.9)^2 = 49$

16. $(x-6)^2 = 81$

17. $(x-12)^2 = 121$

18. $(x+2.5)^2 = 64$

16.5   Completing the Square

"Completing the Square" is another way of factoring a quadratic equation. To complete the square, convert the equation into a perfect square.

Example 9:   Solve $x^2 - 10x + 9 = 0$ by completing the square.

Completing the square:

Step 1:   The first step is to get the constant on the other side of the equation. Subtract 9 from both sides:
$$x^2 - 10x + 9 - 9 = -9$$
$$x^2 - 10x = -9$$

Step 2:   Determine the coefficient of the x. The coefficient in this example is -10. Divide the coefficient by 2 and square the result.
$$(-10 \div 2)^2 = (-5)^2 = 25$$

Step 3:   Add the resulting value, 25, to both sides:
$$x^2 - 10x + 25 = -9 + 25$$
$$x^2 - 10x + 25 = 16$$

Step 4:   Now factor the $x^2 - 10x + 25$ into a perfect square:
$$(x - 5)^2 = 16$$

Solving the perfect square:

Step 5:   Take the square root of both sides.
$$\sqrt{(x-5)^2} = \sqrt{16}$$
$$(x - 5) = \pm 4$$
$$(x - 5) = 4 \text{ and } (x - 5) = -4$$

Step 6:   Solve the two equations.
$$\begin{array}{lll} (x-5) = 4 & \text{and} & (x-5) = -4 \\ x - 5 + 5 = 4 + 5 & \text{and} & x - 5 + 5 = -4 + 5 \\ x = 9 & \text{and} & x = 1 \end{array}$$

Solve for x by completing the square.

1. $x^2 + 2x - 3 = 0$

2. $x^2 - 8x + 7 = 0$

3. $x^2 + 6x - 7 = 0$

4. $x^2 - 16x - 36 = 0$

5. $x^2 - 14x + 49 = 0$

6. $x^2 - 4x = 0$

7. $x^2 + 12x + 27 = 0$

8. $x^2 + 2x - 24 = 0$

9. $x^2 + 12x - 85 = 0$

10. $x^2 - 8x + 15 = 0$

11. $x^2 - 16x + 60 = 0$

12. $x^2 - 8x - 48 = 0$

13. $x^2 + 24x + 44 = 0$

14. $x^2 + 6x + 5 = 0$

15. $x^2 - 11x + 5.25 = 0$

16.6   Discriminant

The **discriminant**, D, is determined from the coefficients of the quadratic equation $ax^2 + bx + c = 0$. The formula for the discriminant is shown below.

$$D = b^2 - 4ac$$

The discriminant illustrates how many real roots the quadratic equation has. Look at the table below:

Discriminant	Roots
$D < 0$	no real roots (only complex roots)
$D = 0$	one real roots
$D > 0$	two real roots

Example 10:   Find the discriminant, D, of the equation $x^2 + 7x + 12 = 0$.

Step 1:   Determine what a, b, and c are in the quadratic equation $x^2 + 7x + 12 = 0$.
$a = 1$, $b = 7$, $c = 12$

Step 2:   Plus the values for a, b, and c into the formula for the discriminant.
$D = b^2 - 4ac = 7^2 - 4(1)(12) = 49 - 48 = 1$

Step 3:   The discriminant is 1. Looking at the chart above, we see that $1 > 0$, so there are two real roots of the quadratic equation $x^2 + 7x + 12 = 0$.

NOTE: $x^2 + 7x + 12 = 0$ factors to $(x + 3)(x + 4) = 0$

Solve for x:
$$\begin{array}{cccc} x + 3 & = & 0 & \quad x + 4 & = & 0 \\ x & = & -3 & \quad x & = & -4 \end{array}$$

We see that the roots are -4 and -3. Therefore, there are two real roots for the quadratic equation $x^2 + 7x + 12 = 0$.

Find the discriminant of the quadratic equations.

1. $x^2 + 4x + 4 = 0$
2. $x^2 + 8x - 33 = 0$
3. $x^2 - 9 = 0$
4. $2x^2 + 3x - 14 = 0$

5. $x^2 - 11x + 30 = 0$
6. $2x^2 + 5x - 6 = 0$
7. $x^2 - 4x + 2 = 0$
8. $5x^2 - 6x + 21 = 0$

9. $x^2 + 9x + 14 = 0$
10. $x^2 - 3x + 15 = 0$
11. $x^2 - 6x + 9 = 0$
12. $3x^2 + x + 6 = 0$

First, find the discriminant. Then, look at the chart above to determine how many roots the equations have.

13. $x^2 + 8x + 16 = 0$
14. $3x^2 - x + 2 = 0$
15. $x^2 - 7x + 12 = 0$

16. $x^2 + 8x + 15 = 0$
17. $2x^2 - 20x + 50 = 0$
18. $x^2 - 10x + 24 = 0$

16.7   Graphing Quadratic Equations

Equations that you may encounter on the GA Accelerated Math I EOC test will possibly involve variables which are squared (raised to the second power). The best way to find values for the x and y variables in an equation is to plug one number into x, and then find the corresponding value for y. Then, plot the points and draw a line through the points.

Example 11:   Graph $y = x^2$.

Step 1:   Make a table and find several values for x and y.

x	y
-2	4
-1	1
0	0
1	1
2	4

Step 2:   Plot the points, and draw a curve through the points. Notice the shape of the curve. This type of curve is called a **parabola**. Equations with one squared term will be parabolas.

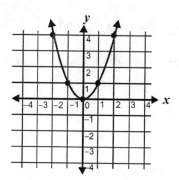

Note: In the equation $y = ax^2 + c$, changing the value of a will widen or narrow the parabola around the y-axis. If the value of a is a negative number, the parabola will be reflected across the x-axis (the vertex will be at the top of the parabola instead of at the bottom.) If $a = 0$, the graph will be a straight line, not a parabola. Changing the value of c will move the vertex of the parabola from the origin to a different point on the y-axis.

Graph the equations below on a Cartesian plane.

1. $y = 2x^2$
2. $y = 3 - x^2$
3. $y = x^2 - 2$
4. $y = -2x^2$
5. $y = x^2 + 3$
6. $y = -3x^2 + 2$
7. $y = 3x^2 - 5$
8. $y = x^2 + 1$
9. $y = -x^2 - 6$
10. $y = -x^2$
11. $y = 2x^2 - 1$
12. $y = 2 - 2x^2$

16.8   Finding the x-Intercepts of a Quadratic Equation

A **quadratic equation** is an equation where either the y or x variable is squared. Finding the intercepts of a quadratic equation is similar to finding the intercepts of a line. In most cases, the variable x is squared, which means there could be two x-intercepts. There could be one, two, or zero x-intercepts. The x-intercepts of a quadratic equation are called the **zeros** or **solutions** of a quadratic equation.

Example 12:   Find the x-intercept(s) of the quadratic equation, $y = x^2 - 4$.

Step 1:   Find the x-intercept(s). The x-intercept is the point, or points in this case, where the graph crosses the x-axis. In this case, we plug 0 in for y because y is always zero along the x-axis. Solve for x.

$$y = x^2 - 4$$
$$0 = x^2 - 4$$
$$0 + 4 = x^2 - 4 + 4$$
$$4 = x^2$$
$$\sqrt{4} = \sqrt{x^2}$$
$$\sqrt{4} = x$$

Thus, $x = \sqrt{4}$ and $\sqrt{4} = -2$ or 2, so $x = -2$ or $x = 2$.
The x-intercepts are $(-2, 0)$ and $(2, 0)$.

Step 2:   To verify that the intercepts are correct, graph the equation on the coordinate plane.

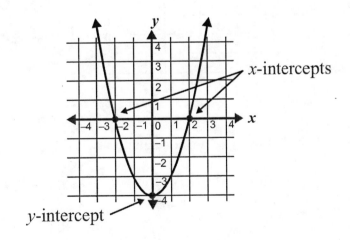

Example 13:   Find the x-intercept(s) of the quadratic equation, $y = x^2 + 2$.

Step 1:   Find the x-intercept. Plug 0 in for y and solve for x.
$$y = x^2 + 2$$
$$0 = x^2 + 2$$
$$0 - 2 = x^2 + 2 - 2$$
$$-2 = x^2$$
$$\sqrt{-2} = \sqrt{x^2}$$
$$\sqrt{-2} = x$$
You cannot take the square root of a negative number and get a real number as the answer, so there is no x-intercept for this quadratic equation.

Step 3:   To verify that the intercepts are correct, graph the equation on the coordinate plane. As you can see from the graph, the tails of the parabola are increasing in the positive y direction, so they will never come back down, which means they will never cross the x-axis.

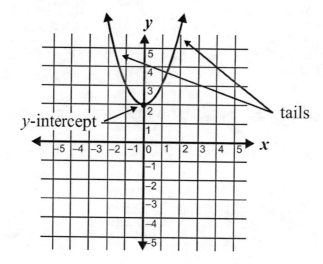

Find the x-intercept(s) of the following quadratic equations.

1. $y = x^2$

2. $y = 2x^2 - 4$

3. $y = -x^2 + 2$

4. $y = x^2 + 6$

5. $y = -x^2 - 1$

6. $y = x^2 - 5$

7. $y = 4x^2 + 8$

8. $y = x^2 + 2x + 1$

9. $y = x^2 - 7x + 12$

16.9   Solving Quadratic Inequalities

A **quadratic inequality** is a polynomial inequality of degree 2. The standard form of a quadratic inequality is similar to that of a quadratic equation, which has the standard form $ax^2 + bx + c = 0$, where $a \neq 0$. The only difference in the standard form of a quadratic inequality is that the equal sign in the standard form of a quadratic equation is replaced by either $>$, $\geq$, $<$, or $\leq$.

Example 14:   Solve the inequality $-x(x - 2) < -3$.

Step 1:   First put all of the terms in the quadratic inequality on one side of the inequality sign.
$$-x(x - 2) < -3 \rightarrow -x^2 + 2x + 3 < 0$$

Step 2:   Now find the root(s) of the corresponding quadratic equation.
$$(x - 3)(-x - 1) = 0$$
$$x - 3 = 0 \qquad -x - 1 = 0 \qquad \text{Set each factor equal to zero.}$$
$$x = 3 \qquad\qquad x = -1$$

Step 3:   Graphing the two points -1 and 3 on a number line determines the three regions that could satisfy the inequality.

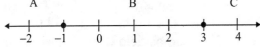

Step 4:   Now, pick a number from each of three regions in the graph above to determine which region satisfies the inequality.

Region	Number Chosen	Substitute into Equation	Satisfy Inequality?
A $(x < -1)$	-2	$-(-2)^2 + 2(-2) + 3 = -5 < 0$	Yes
B $(-1 < x < 3)$	0	$-(0)^2 + 2(0) + 3 = 3 > 0$	No
C $(x > 3)$	4	$-(4)^2 + 2(4) + 3 = -5 < 0$	Yes

Step 5:   The numbers from region A and C satisfy the inequality. The solution to the quadratic inequality is $x < -1$ or $x > 3$. A graph of this solution is shown below.

Solve each of the following quadratic inequalities.

1. $x^2 + 2x - 8 > 0$

2. $x(x + 7) \geq -12$

3. $-x^2 - 5x + 14 < 0$

4. $5x^2 + 34x - 7 \leq 0$

5. $x^2 - 14x + 24 \geq 0$

6. $x(x + 8) \leq 9$

7. $x^2 - 11x + 30 > 0$

8. $3x^2 + 8x < 16$

9. $-x^2 + x + 56 > 0$

10. $-x^2 + 4x - 9 \leq 0$

11. $x^2 - x + 7 < 0$

12. $x(x - 4) > -5$

13. $4x^2 - x + 1 \geq 0$

14. $(x - 3)^2 < 0$

15. $x^2 + 2x + 6 \leq 0$

16. $5x(x - 1) > -2$

17. $8x^2 - 7x + 2 \geq 0$

18. $-x^2 + 3x - 11 > 0$

16.10   Graphing Quadratic Inequalities

A **quadratic inequality** can also be solved by graphing the inequality on a coordinate plane.

Example 15:   Graph the quadratic inequality $y < -2x^2 + 4$, then find the solution.

Step 1:   Change the inequality to an equality: $y = -2x^2 + 4$. Graph the quadratic equation as you did on page 232.

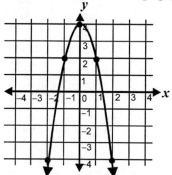

Step 2:   Change the equality back to an inequality: $y < -2x^2 + 4$. This inequality is less than, so it will have a dotted line instead of a solid line. Also, the less than sign implies that there will be shading below the graph instead of above it.

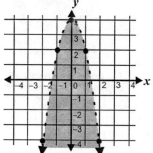

Step 3:   To find the solution for the quadratic inequality, find the x-intercepts first. The x-intercepts are $\left(\pm\sqrt{2}, 0\right)$. Then looking at the graph, we see the shaded portion of the graph is between the two intercepts.
Therefore, the solution of the inequality is $-\sqrt{2} < x < \sqrt{2}$.

Graph the inequalities below on a Cartesian plane, then find the solution(s).

1. $y \leq 2x^2$

2. $y > 3 - x^2$

3. $y \geq x^2 + 2x - 2$

4. $y \leq -3x^2 + 2$

5. $y > x^2 + 1$

6. $y > x^2$

7. $y \geq x^2 + 6x + 1$

8. $y \leq x^2 - 2x + 3$

Chapter 16 Review

Factor and solve each of the following quadratic equations.

1. $16b^2 - 25 = 0$

2. $a^2 - a - 30 = 0$

3. $x^2 - x = 6$

4. $100x^2 - 49 = 0$

5. $81y^2 = 9$

6. $y^2 = 21 - 4y$

7. $y^2 - 7y + 8 = 16$

8. $6x^2 + x - 2 = 0$

9. $3y^2 + y - 2 = 0$

10. $b^2 + 2b - 8 = 0$

11. $4x^2 + 19x - 5 = 0$

12. $8x^2 = 6x + 2$

13. $2y^2 - 6y - 20 = 0$

14. $-6x^2 + 7x - 2 = 0$

15. $y^2 + 3y - 18 = 0$

Using the quadratic formula, find both solutions for the variable.

16. $x^2 + 10x - 11 = 0$

17. $y^2 - 14y + 40 = 0$

18. $b^2 + 9b + 18 = 0$

19. $y^2 - 12y - 13 = 0$

20. $a^2 - 8a - 48 = 0$

21. $x^2 + 2x - 63 = 0$

22. $-3x^2 - 2x - 2 = 0$

23. $4x^2 + x + 5 = 0$

24. $2x^2 - 8x + 9 = 0$

Solve each of the following quadratic inequalities.

25. $2x^2 + x - 15 \geq 0$

26. $-2x^2 + 21x + 11 < 0$

27. $x(3x + 1) \leq 24$

28. $-5x^2 - 9x - 12 > 0$

29. $x^2 + 8x + 30 \geq 0$

30. $6x^2 - 7x + 8 \leq 0$

Find the x-intercept(s) of the following quadratic equations.

31. $y = 2x^2 - 8$

32. $y = -x^2 - 9$

33. $y = 5x^2 + 1$

34. $y = x^2 - 6$

Graph the equations below on a Cartesian plane.

35. $y = -x^2 + 4$

36. $y = 3x^2 - 1$

37. $y = 4 - x^2$

38. $y = 3x^2 + 6$

Graph the inequalities below on a Cartesian plane, then find the solution(s).

39. $y < -x^2 - 6$

40. $y < x^2 + 4x - 1$

41. $y > x^2 + 2$

42. $y > 3x^2 + 4$

Chapter 16 Test

1. Solve: $4y^2 - 9y = -5$

(A) $\left\{ 1, \dfrac{5}{4} \right\}$

(B) $\left\{ -\dfrac{3}{4}, -1 \right\}$

(C) $\left\{ -1, \dfrac{4}{5} \right\}$

(D) $\left\{ \dfrac{5}{16}, 1 \right\}$

2. Solve for y: $2y^2 + 13y + 15 = 0$

(A) $\left\{ \dfrac{3}{2}, \dfrac{5}{2} \right\}$

(B) $\left\{ \dfrac{2}{3}, \dfrac{2}{5} \right\}$

(C) $\left\{ -5, -\dfrac{3}{2} \right\}$

(D) $\left\{ 5, -\dfrac{3}{2} \right\}$

3. Solve for x.

$x^2 - 3x - 18 = 0$

(A) $\{-6, 3\}$
(B) $\{6, -3\}$
(C) $\{-9, 2\}$
(D) $\{9, -2\}$

4. What are the values of x in the quadratic equation?

$x^2 + 2x - 15 = x - 3$

(A) $\{-4, 3\}$
(B) $\{-3, 4\}$
(C) $\{-3, 5\}$
(D) Cannot be determined

5. Solve the equation $(x + 9)^2 = 49$

(A) $x = -9, 9$
(B) $x = -9, 7$
(C) $x = -16, -2$
(D) $x = -7, 7$

6. Solve the equation $c^2 + 8c - 9 = 0$ by completing the square.

(A) $c = \{1, -9\}$
(B) $c = \{-1, 9\}$
(C) $c = \{3, 3\}$
(D) $c = \{-3, -3\}$

7. Using the quadratic formula, solve the following equation:

$3x^2 = 9x$

(A) $x = \{0, 1\}$
(B) $x = \{3, 1\}$
(C) $x = \{0, 3\}$
(D) $x = \{3, -3\}$

8. Solve $6a^2 + 11a - 10 = 0$, using the quadratic formula.

(A) $\left\{ -\frac{2}{5}, \frac{3}{2} \right\}$

(B) $\left\{ \frac{2}{5}, \frac{2}{3} \right\}$

(C) $\left\{ -\frac{5}{2}, \frac{2}{3} \right\}$

(D) $\left\{ \frac{5}{2}, \frac{2}{3} \right\}$

9. Which of the following quadratic inequalities has no real solution?

(A) $3x^2 - 17x - 6 < 0$
(B) $-4x^2 - 11x + 3 > 0$
(C) $5x^2 - 7x - 6 \leq 0$
(D) $-6x^2 - 3x - 8 \geq 0$

10. Which of the following solutions is correct for the quadratic inequality $6x^2 - 41x - 7 \le 0$?

(A) $x \le -6$ or $x \ge 7$

(B) $x \le -\frac{1}{6}$ or $x \ge 7$

(C) $-6 \le x \le 7$

(D) $-\frac{1}{6} \le x \le 7$

11. Which of the following graphs represents $y = 2x^2$?

(A)

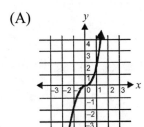

(B)

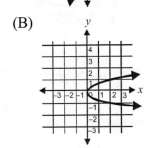

(C)

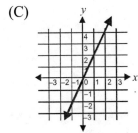

(D)

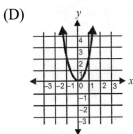

12. Which of the following expressions is a root of the quadratic equation $-7x^2 - x - 3 = 0$?

(A) $-\dfrac{1}{14} - \dfrac{\sqrt{83}}{14}$

(B) $1 + \sqrt{83}i$

(C) $-\dfrac{1}{14} - \dfrac{\sqrt{83}}{14}i$

(D) $1 - \sqrt{83}i$

13. What are the x-intercept(s) of the equation $y = -2x^2 + 8$?

(A) $(-2, 0)$ and $(2, 0)$

(B) $(-4, 0)$ and $(4, 0)$

(C) $(8, 0)$

(D) There are no x-intercepts.

14. What are the x-intercept(s) of the equation $y = x^2 + 9$?

(A) $(9, 0)$

(B) $(-3, 0)$ and $(3, 0)$

(C) $(-9, 0)$ and $(9, 0)$

(D) There are no x-intercepts.

15. Looking at the graph below, what is the solution to the inequality $y \le x^2 - 10x + 19$?

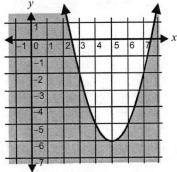

(A) $x = 5 \pm \sqrt{6}$

(B) $x < 5 - \sqrt{6}$ or $x > 5 + \sqrt{6}$

(C) $5 - \sqrt{6} \le x \le 5 + \sqrt{6}$

(D) $x \le 5 - \sqrt{6}$ or $x \ge 5 + \sqrt{6}$

Chapter 17
Quadratic Functions

This chapter covers the following Georgia Performance Standards:

MA1A	Algebra	MA1A3a
		MA1A3b
		MA1A3c
		MA1A3d
		MA1A3e

17.1 Converting Quadratic Functions Between Standard and Vertex Form

The **standard form** of a quadratic function is $f(x) = ax^2 + bx + c$, where a, b, and c are constants, and $a \neq 0$.

The **vertex form** of a quadratic function is $f(x) = a(x-h)^2 + k$, where a, h, and k are constants, and $a \neq 0$. The reason why $f(x) = a(x-h)^2 + k$ is referred to as vertex form is because the point (h, k) is the vertex of the graph of the quadratic function.

Example 1: Convert the quadratic function $f(x) = 5(x+2)^2 - 9$ from vertex form to standard form.

Step 1: To convert the function to standard form, simplify the function using order of operations.

$$f(x) = 5(x+2)^2 - 9$$
$$f(x) = 5(x+2)(x+2) - 9$$
$$f(x) = 5(x^2 + 4x + 4) - 9$$
$$f(x) = 5x^2 + 20x + 20 - 9$$

Step 2: Now, combine like terms.

$$f(x) = 5x^2 + 20x + 20 - 9 = 5x^2 + 20x + 11$$

The standard form of the function $f(x) = 5(x+2)^2 - 9$ is
$f(x) = 5x^2 + 20x + 11$

Convert each of the following quadratic functions from vertex form to standard form.

1. $f(x) = -3(x-4)^2 + 7$

2. $f(x) = \frac{2}{3}(x+1)^2 - 8$

3. $f(x) = 6(x+10)^2 - 2$

4. $f(x) = 4(x-7)^2 + 11$

5. $f(x) = -(x+4)^2 - 3$

6. $f(x) = 7(x-9)^2 + 1$

7. $f(x) = 5(x+3)^2 - 4$

8. $f(x) = 2(x-5)^2 + 9$

9. $f(x) = 9(x+2)^2 - 5$

Example 2:   Convert the quadratic function $f(x) = 5x^2 - 40x + 67$ from standard form to vertex form.

Step 1:   To convert the function to vertex form, the completing the square technique must be used. The first step is to subtract the constant from both sides of the equation:
$$f(x) = 5x^2 - 40x + 67$$
$$f(x) - 67 = 5x^2 - 40x + 67 - 67$$
$$f(x) - 67 = 5x^2 - 40x$$

Step 2:   Next, 5 should be factored from the right side of the equation:
$$f(x) - 67 = 5x^2 - 40x = 5(x^2 - 8x)$$

Step 3:   Next, the square should be completed on the right side of the equation. To review completing, refer to page 230. Divide the coefficient of x by 2, then square the result. $(-8) \div 2 = -4$. $(-4)^2 = 16$. Add 16 to the binomial in the parenthesis.
$$f(x) - 67 = 5(x^2 - 8x)$$
$$f(x) - 67 = 5(x^2 - 8x + 16)$$

Step 4:   Since 16 is added to the right side of the function within the parentheses, it is multiplied by 5. This means that we added $5(16) = 80$ to the right side of the function, not just 16. So we must add 80 to the left side of the function in order for the function to be balanced.
$$f(x) - 67 + 80 = 5(x^2 - 8x + 16)$$

Step 5:   Next, simplify the function by factoring the trinomial on the right and combining like terms on the left.
$$f(x) - 67 + 80 = 5(x^2 - 8x + 16)$$
$$f(x) + 13 = 5(x - 4)^2$$

Step 6:   Finally, 13 should be subtracted from both sides of the equation so that the function is in vertex form.
$$f(x) + 13 = 5(x - 4)^2$$
$$f(x) + 13 - 13 = 5(x - 4)^2 - 13$$
$$f(x) = 5(x - 4)^2 - 13 \text{ is the vertex form of the function.}$$

Convert each of the following quadratic functions from standard form to vertex form.

1. $f(x) = 8x^2 - 16x + 27$ 4. $f(x) = 5x^2 - 10x + 37$ 7. $f(x) = 7x^2 + 28x + 19$

2. $f(x) = -2x^2 - 24x - 75$ 5. $f(x) = 6x^2 + 72x + 228$ 8. $f(x) = -3x^2 - 24x - 154$

3. $f(x) = 3x^2 + 42x + 121$ 6. $f(x) = -4x^2 + 64x - 270$ 9. $f(x) = 9x^2 - 54x - 10$

17.2   Graphing Transformations of the Quadratic Parent Function

The quadratic parent function is $f(x) = x^2$. All quadratic functions can be graphed as a transformation of the parent function. The function must first be put into vertex form, $g(x) = a(x-h)^2 + k$. This allows us to see the transformations more clearly. Four types of transformations can occur:

1. If $a \neq 1$, then horizontal stretching/shrinking of the graph of x^2 occurs. It becomes wider when $0 < a < 1$ or thinner if $a > 1$.

2. If $h \neq 0$, then a translation left or right will occur. Move the graph h units horizontally.

3. If $k \neq 0$, then a translation up or down will occur. Move the graph k units vertically.

4. If a is negative, reflect the graph over the x-axis.

Note: It does not matter which type of transformation happens first.

Example 3:   Graph the quadratic function $f(x) = \frac{1}{3}(x+4)^2 - 10$ as a transformation of the function $f(x) = x^2$.

The graph of $f(x) = x^2$ is graphed below.

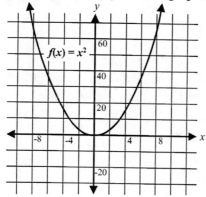

Step 1:   To transform the graph of the function $f(x) = x^2$ into the graph of the function $f(x) = \frac{1}{3}(x+4)^2 - 10$, we will start with horizontal stretching. Since $a = \frac{1}{3} < 1$, we know that the parent function will become wider. When this is done, the function becomes $f(x) = \frac{1}{3}x^2$.

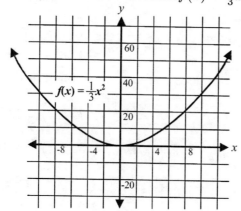

Copyright © American Book Company

Step 2: Next, a translation of 10 units down will be applied, $k = -10$. When this is done, the function becomes $f(x) = \frac{1}{3}x^2 - 10$.

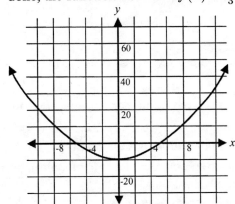

Step 3: Finally, a translation of 4 units left will be applied, $h = -4$. When this is done, the function becomes $f(x) = \frac{1}{3}(x + 4)^2 - 10$.

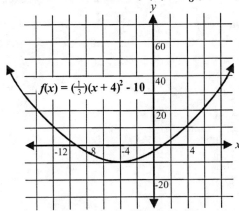

Graph each of the following quadratic functions as transformations of the parent function.

1. $f(x) = 3(x + 2)^2 - 4$

2. $f(x) = \frac{1}{4}(x - 6)^2 - 8$

3. $f(x) = 2(x + 5)^2 + 9$

4. $f(x) = 5(x - 1)^2 + 7$

5. $f(x) = \frac{2}{5}(x + 3)^2 - 1$

6. $f(x) = 4(x - 9)^2 - 3$

7. $f(x) = 7(x + 1)^2 + 5$

8. $f(x) = \frac{2}{3}(x - 2)^2 + 10$

9. $f(x) = 6(x + 4)^2 - 6$

10. $f(x) = 7x^2 + 28x + 18$

11. $f(x) = \frac{1}{2}x^2 - 8x + 44$

12. $f(x) = 5x^2 + 30x + 38$

13. $f(x) = -3x^2 + 12x - 34$

14. $f(x) = 4x^2 + 72x + 307$

15. $f(x) = \frac{1}{3}x^2 - 6x + 33$

16. $f(x) = -6x^2 - 12x + 23$

17. $f(x) = 15x^2 + 2x + 3$

18. $f(x) = 8x^2 - 48x + 57$

19. Explain why the graph of $f(x) = (-(x + 4))^2$ is not the graph of $f(x) = (x + 4)^2$ reflected in the y-axis.

17.3   Characteristics of Quadratic Functions

Every quadratic function has various characteristics.

The **domain** of the quadratic function $f(x)$ is all possible values of x for the function, the **range** is all possible values of $f(x)$.

The **zeros** or **roots** are the values of x that produce a value of 0 for $f(x)$, and the **extrema** is the maximum or minimum value of $f(x)$.

The graph of every quadratic function, which is a parabola, has certain properties as well.

The **axis of symmetry** of the graph of the function $f(x)$ is a line down the middle of the graph that passes through the vertex and divides the parabola into two equal parts.

The **vertex** of the graph is the point of intersection between the axis of symmetry and the parabola. It is the highest or lowest point in the center of a parabola.

The **intercepts** of the graph are the points where the parabola crosses the x- and y-axes.

The **intervals of increase and decrease** of the graph of the function $f(x)$ are the regions in which the parabola is going up or going down.

Example 4:   The graph of the function $f(x) = 3(x + 4)^2 + 6$ is shown. List the function's domain, range, zeros, and extrema, and the graph's axis of symmetry, vertex, intercepts, and intervals of increase/decrease.

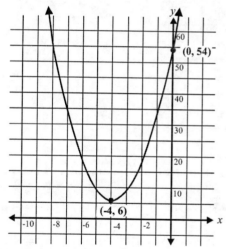

Domain: all real numbers

Range: $f(x) \geq 6$. The range of the function is $6 \leq y \leq \infty$.

Zeros: The function does not cross x-axis, so $f(x)$ never equals zero and the function has no zeros.

Extrema: The function has a minimum at 6.

Symmetry: Since the graph of the function can be reflected over the line $x = -4$ and produce the same graph, its axis of symmetry is $x = -4$.

Vertex: The graph's vertex is $(-4, 6)$. This can be found from the vertex form.

Intercepts: The graph does not have any x-intercepts. The function has a y-intercept at $(0, 54)$.

Intervals: The graph has an interval of decrease of $(-\infty, -4]$ and an interval increase of $[-4, \infty)$.

Example 5:   The graph of the function $f(x) = -5(x-4)^2 + 2$ is shown. List the function's domain, range, zeros, and extrema, and the graph's axis of symmetry, vertex, intercepts, and intervals of increase/decrease.

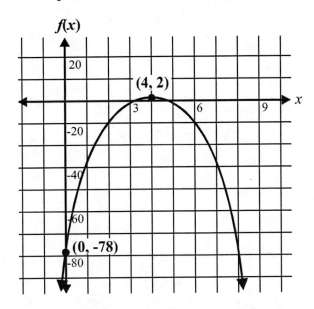

Domain: all real numbers

Range: The range of the function is all values less than or equal to 2.

Zeros: The zeros of the function can be determined by plugging 0 in for $f(x)$ and solving for x.

Extrema: Where $x = 4$, $f(x) = 2$. For all other values of x, $f(x) < 2$. Therefore, the function has a maximum value of 2, the extrema.

Symmetry: The graph of the function can be reflected on the line $x = 4$ and produce the same graph, its axis of symmetry is $x = 4$.

Vertex: The axis of symmetry intersects the parabola at the point $(4, 2)$, the graph's vertex is $(4, 2)$.

Intercepts: The zeros of the function are $x = 4 \pm \frac{\sqrt{10}}{5}$, the graph has x-intercepts at $x = 4 - \frac{\sqrt{10}}{5}$ and $x = 4 + \frac{\sqrt{10}}{5}$. Plugging in zero for x and solving for $f(x)$ shows that the function has a y-intercept at $f(x) = -78$.

Intervals: The graph has an interval of decrease of $[4, \infty)$ and an interval increase of $(-\infty, 4]$.

For each of the following quadratic functions, list the function's domain, range, zeros, and extrema, and for the graph of the function list the axis of symmetry, vertex, intercepts, and intervals of increase and decrease.

1. $f(x) = 2(x - 5)^2 + 18$

7. $f(x) = -2(x + 8)^2 + 5$

2. $f(x) = \frac{1}{4}(x + 9)^2 + 1$

8. $f(x) = 5(x - 3)^2 - 11$

3. $f(x) = 8(x - 2)^2 + 3$

9. $f(x) = -\frac{5}{6}(x + 1)^2 + 2$

4. $f(x) = 9(x - 1)^2 + 10$

10. $f(x) = \frac{7}{2}(x - 2)^2 - 14$

5. $f(x) = 5(x + 6)^2 + 14$

11. $f(x) = 4(x - 4)^2 - 5$

6. $f(x) = \frac{4}{5}(x + 3)^2 + 2$

12. $f(x) = -3(x + 5)^2 + 9$

Example 6:   Determine the equation of the quadratic function whose graph has a vertex of $(5, 8)$ and a y-intercept of $(0, -67)$.

Step 1:   Since the quadratic function has a graph whose vertex is $(5, 8)$, the equation of the function must be $f(x) = a(x - 5)^2 + 8$, where a is a constant. Also, since the y-intercept of the graph is $f(x) = -67$, when 0 is plugged in for x, the value of $f(x)$ must equal -67. For this reason, a can be solved for as follows:

$-67 = a(0 - 5)^2 + 8$
$-67 = a(5)^2 + 8$
$-67 = 25a + 8$
$-67 - 8 = 25a + 8 - 8$
$-75 = 25a$
$a = -3$

Step 2:   Therefore, the equation of the quadratic function must be $f(x) = -3(x - 5)^2 + 8$.

Determine the equation of the quadratic function that has each of the following vertices and y-intercepts.

13. Vertex: $(-1, 7)$; y-intercept:$(0, 12)$

15. Vertex: $(5, 2)$; y-intercept: $(0, 152)$

14. Vertex: $(6, -8)$; y-intercept: $(0, 64)$

16. Vertex: $(-4, -11)$; y-int.: $f(x) = (0, -9)$

17.4   Determining the Rate of Change of Quadratic Functions

The average rate of change for a function is the change in $f(x)$ divided by the change in x. In linear equations, the rate of change (slope) is constant. In quadratic functions, the rate of change is not constant. For any point on a parabola, the rate of change is different. The average rate of change for $f(x)$ from one value of x, x_1, to another, x_2, can be calculated:

$$\frac{f(x_1) - f(x_2)}{x_1 - x_2}$$

By choosing a value of x_2 that is close to x_1 we can approximate the rate of change of f at x_1. Remember, you must keep the order of the terms in the numerator and the denominator consistent. A positive rate of change will have a graph that "goes up." A negative rate of change will have a graph that "goes down." A rate of change of 0 indicates a horizontal line. A very large rate of change indicates a nearly vertical line.

Example 7:   What is the average rate of change for $f(x) = -2(x + 2)^2 + 2$ from $x = -2$ to $x = 2$?

Step 1:   Find $f(-2)$ and $f(2)$ by plugging in $x = -2$ and $x = 2$ into the equation.
$$f(-2) = -2(-2 + 2)^2 + 2 = 2$$
$$f(2) = -2(2 + 2)^2 + 2 = -30$$

Step 2:   Find the difference between $f(2)$ and $f(-2)$. Plug in the values for x.
$$f(2) - f(-2) = -30 - 2 = -32.$$

Step 3:   Find the difference between 2 and -2. $2 - (-2) = 4$

Step 4:   Divide the change in $f(x)$ by the change in x. $-32 \div 4 = -8$
The average rate of change of the function from $x = -2$ to $x = 2$ is -8.

Find average rate of change of each of the following quadratic functions from $x = 2$ to $x = 5$.

1. $f(x) = 7(x - 1)^2 - 3$

2. $f(x) = x^2 - 6x + 19$

3. $f(x) = 9(x - 8)^2 - 5$

4. $f(x) = 4x^2 - 7x - 2$

5. $f(x) = -2(x + 9)^2 + 4$

6. $f(x) = x^2 + 4x - 21$

7. $f(x) = 5(x - 7)^2 + 9$

8. $f(x) = 3x^2 + 36x + 107$

9. $f(x) = -6(x + 1)^2 + 12$

10. $f(x) = x^2 - 3x - 18$

11. $f(x) = 10(x + 2)^2 - 4$

12. $f(x) = -2x^2 - 8x - 21$

17.5   Arithmetic Series

An **arithmetic sequence** is an ordered list of numbers where the difference between each successive number is constant. The sequence $\{1, 2, 3, 4, ...\}$ is an arithmetic sequence because the difference between one member and the next is 1. An **arithmetic series** is the sum of an arithmetic sequence of numbers.

To find any term of an arithmetic series, the formula $a_n = a_1 + d(n-1)$ can be used, where a_1 is the first term, d is the common difference, and n is the number of the term to find.

To calculate the sum of the first n terms of an arithmetic series, use the formula $S_n = \dfrac{n(a_1 + a_n)}{2}$, where a_1 is the first term in the series and a_n is the nth term in the series. This formula will not be provided on your formula sheet.

Example 8:   Find the sum of the first 30 terms of the arithmetic series:
$3 + 4 + 5 + 6 + 7 + \ldots$

Step 1:   First, find the pattern of the series. The pattern of this series is $n + 2$.

Step 2:   Next find all the variables for the summing equation, $S_n = \dfrac{n(a_1 + a_n)}{2}$.
$a_1 = 3$
$n = 30$ (how many terms we will find the sum of)
$a_n = a_{30} = 30 + 2 = 32$

Step 3:   Plug all the values into the equation. $S_{30} = \dfrac{30(3 + 32)}{2} = 525$.
The sum of the arithmetic series is 525.

Example 9:   Calculate the sum of the first 10 natural numbers.

Method 1:   We can manually write out and compute the sum:.
$1 + 2 + 3 + 4 + 5 + 6 + 7 + 8 + 9 + 10 = 55$

Method 2:   We can use the formula to find S_{10}, where $n = 10$, $a_1 = 1$, and $a_{10} = 10$.
$$S_{10} = \frac{n(a_1 + a_{10})}{2} = \frac{10(1 + 10)}{2} = 55$$

Example 10:   What is the sum of the arithmetic series $\sum_{n=1}^{6} [2 + 5(n-1)]$?

Step 1:   Write out the numbers in the arithmetic series.

The first number in the arithmetic series is $2 + 5(1-1) = 2 + 5(0) = 2 + 0 = 2$, while the second number in the arithmetic series is $2 + 5(2-1) = 2 + 5(1) = 2 + 5 = 7$. For this reason, the common difference of the arithmetic series must be $7 - 2$, or 5, and since there are 6 terms in the arithmetic series, it must be $2 + 7 + 12 + 17 + 22 + 27$.

Step 2:   Add the numbers in the arithmetic series together.
$2 + 7 + 12 + 17 + 22 + 27 = 87$

Find the sum of the first 20 terms in the finite arithmetic series.

1. $5 + 6 + 7 + 8 + 9 + \ldots$

4. $4 + 6 + 8 + 10 + 12 + \ldots$

2. $-3 + -6 + -9 + -12 + -15 + \ldots$

5. $-1 + 0 + 1 + 2 + 3 + \ldots$

3. $0 + -1 + -2 + -3 + -4 + \ldots$

6. $-2 + -1 + 0 + 1 + 2 + \ldots$

Calculate the sum of the arithmetic series with each of the following descriptions.

7. $a_1 = 12$, $a_{16} = 117$, $n = 16$

10. $a_1 = 150$, $a_{25} = -354$, $n = 25$

8. $a_1 = -13$, $a_{22} = 92$; $n = 22$

11. $a_1 = -11$, $a_{31} = 259$, $n = 31$

9. $a_1 = 4$, $a_{19} = 202$, $n = 19$

12. $a_1 = 7$, $a_{51} = -293$, $n = 51$

Calculate the sum of each of the following groups of numbers.

13. the first 23 positive even integers

16. the first 31 multiples of 5

14. the first 18 multiples of 3

17. the first 100 positive even integers

15. the first 44 positive odd integers

18. the first 29 multiples of 4

Calculate the sum of each of the following arithmetic series.

19. $\sum_{n=1}^{5} [-3 + 4(n - 1)]$

22. $\sum_{n=1}^{5} [-11 + 5(n - 1)]$

20. $\sum_{n=1}^{6} [6 - 2(n - 1)]$

23. $\sum_{n=1}^{6} [13 - 3(n - 1)]$

21. $\sum_{n=1}^{4} [25 - 7(n - 1)]$

24. $\sum_{n=1}^{4} [-8 + 2(n - 1)]$

17.6   Exploring Sequences in an Arithmetic Series

If you list the partial sums of an arithmetic series repeatedly, you will get a sequence. This sequence can often be described using quadratic functions.

The rate of change of a quadratic function changes at a constant rate. For example, four points on the graph of the function $f(x) = x^2$ are $(1, 1), (2, 4), (3, 9)$, and $(4, 16)$.

Taking the differences between each $f(x)$ value and the previous $f(x)$ value, we get $4 - 1 = 3$, $9 - 4 = 5$, and $16 - 9 = 7$.

The first change in $f(x)$ is 3 units, the second change is 5 units, and the third change is 7 units. For this reason, for every change of 1 unit of x, the change in $f(x)$ is increasing by 2 units each time.

The sequences of partial sums of arithmetic series work in the same way, and are therefore examples of quadratic functions.

Example 11:   Calculate the first differences and the second differences of the sequence of partial sums of the arithmetic series $3 + 8 + 13 + 18 + 23 + 28 + 33$.

Step 1:   Determine the sequence of partial sums of the arithmetic series.

The partial sums of the arithmetic series are as follows:

$S_1 = 3$

$S_2 = 3 + 8 = 11$

$S_3 = 3 + 8 + 13 = 24$

$S_4 = 3 + 8 + 13 + 18 = 42$

$S_5 = 3 + 8 + 13 + 18 + 23 = 65$

$S_6 = 3 + 8 + 13 + 18 + 23 + 28 = 93$

$S_7 = 3 + 8 + 13 + 18 + 23 + 28 + 33 = 126$

Step 2:   Calculate the first differences of the sequence of partial sums of the arithmetic series.

The sequence of partial sums of the arithmetic series is 3, 11, 24, 42, 65, 93, 126. The first differences of the sequence can be calculated by subtracting each partial sum from the partial sum that follows it.

$11 - 3 = 8$

$24 - 11 = 13$

$42 - 24 = 18$

$65 - 42 = 23$

$93 - 65 = 28$

$126 - 93 = 33$

Step 3:   Calculate the second differences of the sequence of partial sums of the arithmetic series.

The sequence of first differences of the partial sums of the arithmetic series is 8, 13, 18, 23, 28, 33. The second differences of the sequence can be calculated by subtracting each first difference from the first difference that follows it.

$$13 - 8 = 5 \qquad 28 - 23 = 5$$
$$18 - 13 = 5 \qquad 33 - 28 = 5$$
$$23 - 18 = 5$$

The second differences are all 5. When the second differences are all the same, it tells us that the polynomial for this sequence of the partial sums of an arithmetic series of values is a quadratic function.

Calculate the first differences and the second differences of the sequence of partial sums of each of the following arithmetic series.

1. $6 + 9 + 12 + 15 + 18$

2. $5 + 11 + 17 + 23 + 29$

3. $4 + 8 + 12 + 16 + 20$

4. $-8 + 0 + 8 + 16 + 24$

5. $9 + 7 + 5 + 3 + 1$

6. $54 + 42 + 30 + 18 + 6$

7. $7 + 12 + 17 + 22 + 27$

8. $13 + 10 + 7 + 4 + 1$

9. $1 + 10 + 19 + 28 + 37$

Example 12:   The sequence of first differences of the partial sums of a series is 2, 4, 8, 16, 32. Determine if the series is arithmetic.

Step 1:   Calculate the second differences of the sequence of partial sums of the series.

The second differences of the sequence can be calculated by subtracting each first difference from the first difference that follows it. Therefore, the second differences of the sequence are as follows:

$$4 - 2 = 2 \qquad 16 - 8 = 8$$
$$8 - 4 = 4 \qquad 32 - 16 = 16$$

Step 2:   Determine if the second differences are all the same. If so, the series is arithmetic. If not, the series is not arithmetic.

The second differences are not all the same. Therefore, the series is not arithmetic.

Determine if the series with each of the following sequences of first differences of partial sums is arithmetic.

10. $11, 19, 27, 35, 43$

11. $3, 9, 27, 81, 243$

12. $1, 2, 3, 5, 8$

13. $5, 12, 19, 26, 33$

14. $2, 10, 50, 250, 1250$

15. $9, 14, 19, 24, 29$

Chapter 17 Review

Convert each of the following quadratic functions from vertex form to standard form.

1. $f(x) = 16(x - 1)^2 + 2$

2. $f(x) = \frac{4}{5}(x + 5)^2 - 9$

Convert each of the following quadratic functions from standard form to vertex form.

3. $f(x) = 6x^2 - 60x + 157$

4. $f(x) = \frac{1}{2}x^2 + 9x + \frac{61}{2}$

5. $f(x) = -3x^2 + 48x - 222$

6. $f(x) = 4x^2 - 24x + 87$

Graph each of the following quadratic functions as transformations of $f(x) = x^2$.

7. $f(x) = -5(x + 3)^2 - 7$

8. $f(x) = 8(x - 10)^2 - 1$

9. $f(x) = \frac{1}{4}(x + 6)^2 + 9$

10. $f(x) = 6x^2 - 24x + 1$

11. $f(x) = 2x^2 + 32x + 123$

12. $f(x) = -\frac{1}{2}x^2 + 7x - \frac{75}{2}$

For each of the following quadratic functions, list the function's domain, range, zeros, and extrema, and for the graph of the function list the axis of symmetry, vertex, intercepts, and intervals of increase and decrease.

13. $f(x) = 4(x - 28)^2 + 19$

14. $f(x) = \frac{1}{8}(x + 11)^2 + 9$

15. $f(x) = 14(x - 3)^2 + 1$

16. $f(x) = -8(x + 5)^2 + 28$

17. $f(x) = \frac{9}{2}(x - 10)^2 - 12$

18. $f(x) = 15(x + 7)^2 - 6$

Find the average rate of change of each of the following quadratic functions from $x = 1$ to $x = 5$.

19. $f(x) = 5(x - 12)^2 - 2$

20. $f(x) = -4(x + 9)^2 + 6$

21. $f(x) = 3x^2 + 12x - 9$

22. $f(x) = \frac{1}{5}x^2 - 4x + 13$

Calculate the sum of each of the following arithmetic series.

23. $\sum_{n=1}^{8} [10 - 3(n - 1)]$

24. $\sum_{n=1}^{7} [14 + 4(n - 1)]$

Find the sum of the finite series when $n = 10$.

25. $0 + 1 + 2 + 3 + 4 + \ldots$

27. $0 + -1 + -2 + -3 + -4 + \ldots$

26. $3 + 6 + 9 + 12 + 15 + \ldots$

28. $2 + 5 + 8 + 11 + 14 + \ldots$

Calculate the sum of the arithmetic series with each of the following descriptions. Use the formula $S_n = \dfrac{n(a_1 + a_n)}{2}$.

29. $a_1 = 5$, $a_{20} = 81$; $n = 20$

30. $a_1 = \frac{3}{2}$, $a_{15} = \frac{45}{2}$; $n = 15$

Calculate the sum of each of the following groups of numbers.

31. the first 75 positive odd integers

32. the first 59 multiples of 6

Calculate the first differences and the second differences of the sequence of partial sums of each of the following arithmetic series.

33. $29 + 26 + 23 + 20 + 17$

35. $-52 + 12 + 72 + 132 + 192$

34. $2 + 11 + 20 + 29 + 38$

36. $21 + 17 + 13 + 9 + 5$

Determine if the series with each of the following sequences of first differences of partial sums is arithmetic.

37. $4, 16, 64, 256, 1024$

38. $13, 20, 27, 34, 41$

Chapter 17 Test

1. Which of the following arithmetic series has a sum of 105?

   (A) $\sum_{n=1}^{7} [5 + 2(n - 1)]$

   (B) $\sum_{n=1}^{7} [7 + 2(n - 1)]$

   (C) $\sum_{n=1}^{7} [9 + 2(n - 1)]$

   (D) $\sum_{n=1}^{7} [11 + 2(n - 1)]$

2. Which of the following quadratic functions has a maximum?

   (A) $f(x) = -(x + 8)^2 + 9$
   (B) $f(x) = \frac{3}{7}(x - 7)^2 + 3$
   (C) $f(x) = (x + 8)^2 + 9$
   (D) $f(x) = \frac{7}{3}(x + 7)^2 - 3$

3. Which transformation can be applied to the graph of the quadratic function $f(x) = x^2$ to produce the graph of the function $f(x) = (x + 9)^2$?

   (A) A horizontal translation of 9 units left.
   (B) A horizontal translation of 9 units right.
   (C) A vertical translation of 9 units down.
   (D) A vertical translation of 9 units up.

4. Today Mickey began collecting baseball cards by purchasing 1 pack of 20 cards. For each of the next 6 days, he plans on purchasing 3 more packs than he purchased the day before. How many baseball cards will Mickey have after the next 6 days?

   (A) 700
   (B) 1400
   (C) 2100
   (D) 2800

5. Which of the following quadratic equations can be solved with the graph shown?

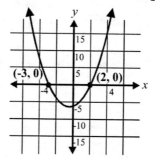

   (A) $x^2 - x - 6 = 0$
   (B) $x^2 - x + 6 = 0$
   (C) $x^2 + x - 6 = 0$
   (D) $x^2 + x + 6 = 0$

6. The table below shows the partial sums of a series, with n being the numbers of terms and S_n being the partial sums. Is the series arithmetic?

n	S_n
1	4
2	16
3	36
4	64

   (A) Yes, because the second differences of the partial sums are each 4.
   (B) Yes, because the second differences of the partial sums are each 6.
   (C) Yes, because the second differences of the partial sums are each 8.
   (D) No, because the second differences of the partial sums are not constant.

7. What is $f(x) = 8x^2 - 48x + 58$ in vertex form?

   (A) $f(x) = 8(x - 3)^2 - 14$
   (B) $f(x) = 8(x - 3)^2 + 14$
   (C) $f(x) = 8(x + 3)^2 - 14$
   (D) $f(x) = 8(x + 3)^2 + 14$

8. What is the average rate of change of the quadratic function $f(x) = 5(x + 4)^2 - 16$ from $x = 1$ to $x = 3$?

(A) 30
(B) 40
(C) 50
(D) 60

9. What is the y-intercept of the quadratic function $f(x) = \frac{5}{9}(x - 7)^2 + 4$?

(A) $\left(0, \dfrac{245}{9}\right)$

(B) $\left(0, \dfrac{281}{9}\right)$

(C) $(0, 245)$

(D) $(0, 281)$

10. Which transformation can be applied to the graph of the quadratic function $f(x) = x^2$ to produce the graph of the function $f(x) = 12x^2$?

(A) A horizontal compression by a factor of $\frac{1}{12}$.

(B) A horizontal stretch by a factor of 12.

(C) A vertical compression by a factor of $\frac{1}{12}$.

(D) A vertical stretch by a factor of 12.

11. The partial sums of an arithmetic series are plotted on a coordinate grid, with the x-coordinate of each point being the number of terms and the y-coordinate of each point being the partial sum. If the points are connected, which of the following is formed?

(A) part of a circle
(B) part of an ellipse
(C) part of a hyperbola
(D) part of a parabola

12. The graph of the quadratic function $f(x) = x^2 + 4x + 7$ is shown. Based on this information, how many solutions does the quadratic equation $x^2 + 4x = -7$ have?

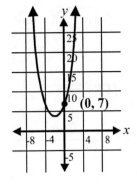

(A) 0
(B) 1
(C) 2
(D) It cannot be determined.

13. The graph of the function $f(x) = x^2$ has been horizontally and vertically translated so that its new vertex is the point $(-5, -11)$. What is the equation of the function with the resulting graph in standard form?

(A) $f(x) = x^2 - 10x + 14$
(B) $f(x) = x^2 - 10x + 36$
(C) $f(x) = x^2 + 10x + 14$
(D) $f(x) = x^2 + 10x + 36$

Chapter 18
Step and Piecewise Functions

This chapter covers the following Georgia Performance Standards:

MA1A	Algebra	MA1A5a
		MA1A5b
		MA1A5c

18.1   Graphs of Step and Piecewise Functions

There are a number of types of graphs that you may see on the GA Math I which are not linear or quadratic.

A **piecewise** function is a function consisting of 2 or more formulas over a sequence of intervals. These intervals are defined by the possible values of x, also known as the domain of the function. The graph of a piecewise function consists of the graphs of each interval formula.

Example 1: $f(x) = \begin{cases} x & \text{where} \quad x < -2 \\ 2x + 1 & \text{where} \quad x \geq -2 \end{cases}$

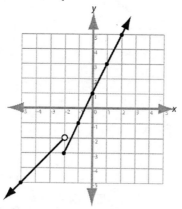

Notice that there is a break in the graph at $x = -2$. Whenever you would have to pick up your pencil in order to draw the graph from left to right, the function is **discontinuous**. At every other point in a function, the function is **continuous**.

A particular type of piecewise defined function is the **absolute value** function. The absolute value of a number is its distance from 0. The absolute value of x is written $|x|$.

Example 2: $f(x) = |x| = \begin{cases} -x & \text{where} \quad x < 0 \\ x & \text{where} \quad x \geq 0 \end{cases}$

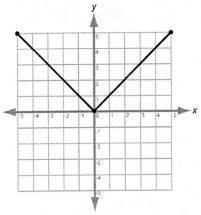

The absolute value function always gives a positive number, but it is **not** discontinuous at $x = 0$, because you do not have to pick up your pencil to draw it.

Another common type of piecewise defined function is a **step** function. Instead of a continuous increase or decrease of y from left to right, the y value stays the same for certain interval of x, then jumps to a higher or lower level. It therefore looks like a set of steps going up or down.

Example 3: $f(x) = \begin{cases} 0 & \text{where} \quad 0 \leq x < 1 \\ 1 & \text{where} \quad 1 \leq x < 2 \\ 2 & \text{where} \quad 2 \leq x < 3 \\ 3 & \text{where} \quad 3 \leq x < 4 \\ 4 & \text{where} \quad 4 \leq x \leq 5 \end{cases}$

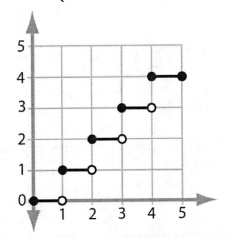

The graph is discontinuous at every "step," in this case $x = 1$, $x = 2$, $x = 3$, and $x = 4$.

18.2   Characteristics of Piecewise Functions

In addition to continuity, there are a number of features that you can identify from the graph of a piecewise function.

Zeros are the points where $y = 0$. The zeros are the function's x-intercepts.

The **maximum** (plural: maxima) is the highest y value that a function reaches.

The **minimum** (plural: minima) is the lowest y value that a function reaches.

The **range** is all of the y values that are defined by the function.

Similarly, the **domain** is all of the x values defined by the function.

Finally, you can determine the **rate of increase or decrease** between two points by finding the rise over the run (slope).

Example 4:   What are the zeroes, maximum, minimum, range, and domain of the function below? At what point is the function discontinuous? Also, what is the rate of increase between $x = 0$ and $x = 2$.

$$y = \begin{cases} 2x & \text{where } 0 \leq x \leq 2 \\ \frac{x}{2} & \text{where } 2 < x \leq 4 \end{cases}$$

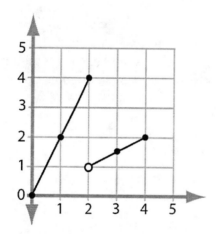

Zeroes: Where does $y = 0$? Where does the graph intersect the x-intercept? Only at $x = 0$.

Maximum: Highest point is $(2, 4)$. Since the maximum point of the graph is always the largest y value that the graph reaches, 4 is the maximum.

Minimum: Lowest point is $(0, 0)$. Since the minimum point of the graph is always the smallest y value that the graph reaches, 0 is the minimum.

Range: Maximum $-$ Minimum $= 4 - 0 = 4$

Domain: Highest x $-$ Lowest $x = 4 - 0 = 4$

The function is discontinuous at $x = 2$.

Rate of increase between $x = 0$ and $x = 2$: At $x = 0$, $y = 0$. At $x = 2$, $y = 4$. Therefore, the rise is $4 - 0 = 4$, and the run is $2 - 0 = 2$.

Slope = rise over run $= \frac{4}{2} = 2$.

Use the following graph of the absolute value function $y = |x - 2|$ for questions 1 and 2.

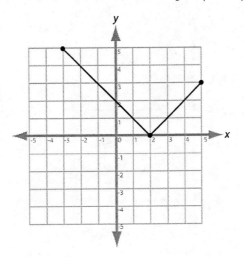

1. According to the graph, what is the minimum value of the function?

   (A) -3
   (B) 0
   (C) 2
   (D) 5

2. At what point is this graph discontinuous?

   (A) $(-3, 5)$
   (B) $(0, 2)$
   (C) $(2, 0)$
   (D) None of the above. The graph is continuous.

Use the following graph of a piecewise defined function for questions 3 and 4.

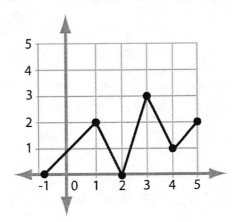

3. How many zeroes does this function have?

   (A) -1
   (B) 0
   (C) 1
   (D) 2

4. What are the range and domain of the function?

   (A) range: $0 \le y \le 3$, domain: $-3 \le x \le 5$
   (B) range: $-3 \le x \le 3$, domain: $0 \le y \le 5$
   (C) range: $-1 \le x \le 5$, domain: $0 \le y \le 3$
   (D) range: $0 \le y \le 3$, domain: $-1 \le x \le 5$

Use the following graph of a step function for questions 5 and 6.

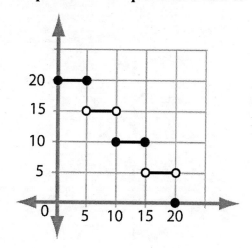

5. At how many points is this graph discontinuous?
   - (A) 4
   - (B) 8
   - (C) 6
   - (D) 2

6. What is the rate of increase of the function between 10 and 15?
   - (A) −5
   - (B) 0
   - (C) 1
   - (D) 5

Use the following graph of a step function for questions 7 and 8.

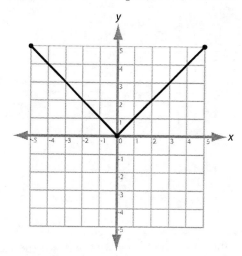

7. What is the vertex of this graph?
   - (A) $(0, 0)$
   - (B) $(-5, 5)$
   - (C) $(5, 5)$
   - (D) $(0, 5)$

8. What is the domain of this function?
   - (A) $-5 \leq x < 5$
   - (B) $0 \leq y < \infty$
   - (C) $0 \leq x < \infty$
   - (D) all real numbers

18.3   More Piecewise Functions

A **piecewise function** is a function consisting of 2 or more formulas over a sequence of intervals. These **intervals** are defined by the possible values of x, also known as the domain of the function. The graph of a piecewise function consists of the graphs of each interval formula.

Example 5:
$$f(x) = \begin{cases} 3 & \text{if } 0 \le x < 1 \\ 2 & \text{if } 1 \le x < 2 \\ 1 & \text{if } 2 \le x < 3 \end{cases}$$

Graph $f(x)$.

Step 1:   Graph each formula over the given interval.

For example, $f(x) = 3$ when the domain is $0 \le x < 1$. This means that you would draw the graph $y = 3$ first. (Recall that this is a horizontal line segment that passes through the point $(0, 3)$.) After this, you would only draw $y = 3$ between the points $(0, 3)$ and $(1, 3)$ because of the domain. The graph cannot go outside of those points.

When $f(x) = 2$ and the domain is $1 \le x < 2$, draw the graph $y = 2$ between the points $(1, 2)$ and $(2, 2)$.

When $f(x) = 1$ and the domain is $2 \le x < 3$, draw the graph $y = 1$ between the points $(2, 1)$ and $(3, 1)$.

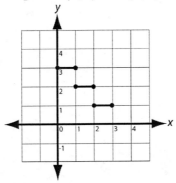

Step 2:   Since the function cannot have two y values for an x value (otherwise, it would not be a function), you must look at the inequalities in the domain. When the inequality is less than or equal to ($\le$), you must draw the endpoint as a filled in circle. This shows that the function includes that point. For the strict inequalities ($<$), you must draw an endpoint with an open (not filled in) circle.

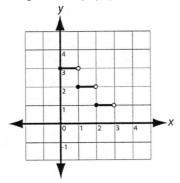

Example 6: $f(x) = \begin{cases} x^2 & \text{if } x \geq 2 \\ 3 - x & \text{if } x < 2 \end{cases}$

Find (A) $f(1)$, (B) $f(3)$, and (C) $f(2)$.

Step 1:   Determine which interval of the domain includes the value of x.
(A) For $f(1)$, $x = 1$. Since 1 is less than 2, you would plug $x = 1$ into $3 - x$.
(B) For $f(3)$, $x = 3$. Since 3 is greater than 2, you would plug $x = 3$ into x^2.
(C) For $f(2)$, $x = 2$. Since 2 is equal to 2, you would plug $x = 2$ into x^2.

Step 2:   Plug the value of x into the appropriate formula to solve for the value of $f(x)$.
(A) $f(x) = 3 - x$, so $f(1) = 3 - 1 = 2$
(B) $f(x) = x^2$, so $f(3) = (3)^2 = 9$
(C) $f(x) = x^2$, so $f(2) = (2)^2 = 4$

Graph each of the following functions.

1. $f(x) = \begin{cases} x & \text{if } x \geq 0 \\ -x & \text{if } x < 0 \end{cases}$

2. $f(x) = \begin{cases} 1 & \text{if } x < 1 \\ x^2 & \text{if } x \geq 1 \end{cases}$

3. $f(x) = \begin{cases} \sqrt{x} & \text{if } x \geq 2 \\ x^2 & \text{if } x < 2 \end{cases}$

4. $f(x) = \begin{cases} 2x + 3 & \text{if } x < 0 \\ 2x - 3 & \text{if } x \geq 0 \end{cases}$

5. $f(x) = \begin{cases} x^2 & \text{if } x < -1 \\ x & \text{if } -1 \leq x \leq 1 \\ -(x^2) & \text{if } x > 1 \end{cases}$

6. Phil's long distance phone service charges him 50 cents for the first 10 minutes and 10 cents for each minute afterwards. Graph the function that represents Phil's long distance phone service and find how much he would pay for

(A) a 5 minute call.
(B) a 10 minute call.
(C) a 15 minute call.

7. The tuition at State University is determined by the number of class hours a student takes. Tuition is $100 for the first three hours and doubles every 3 hours up to 12 hours. After 12 hours, tuition does not change. Graph the function that represents the tuition at State University and find the tuition for a student taking

(A) 6 class hours.
(B) 12 class hours.
(C) 15 class hours.

18.4   Solving Equations and Inequalities with Absolute Values

When solving equations and inequalities which involve variables placed in absolute values, remember that there will be two or more numbers that will work as correct answers. This is because the absolute value variable will signify both positive and negative numbers as answers.

Example 7: $5 + 3|k| = 8$ Solve as you would any equation.

    Step 1: $3|k| = 3$ Subtract 5 from each side.

    Step 2: $|k| = 1$ Divide by 3 on each side.

    Step 3: $k = 1$ or $k = -1$ Because k is an absolute value, the answer can be 1 or -1.

Example 8: $2|x| - 3 < 7$ Solve as you normally would an inequality.

    Step 1: $2|x| < 10$ Add 3 to both sides.

    Step 2: $|x| < 5$ Divide by 2 on each side.

    Step 3: $x < 5$ or $x > -5$ Because x is an absolute value, the answer is a set of both
           or $-5 < x < 5$ positive and negative numbers.

Read each problem, and write the number or set of numbers which solves each equation or inequality.

1. $7 + 2|y| = 15$

2. $4|x| - 9 < 3$

3. $6|k| + 2 = 14$

4. $10 - 4|n| > -14$

5. $-3 = 5|z| + 12$

6. $-4 + 7|m| < 10$

7. $5|x| - 12 > 13$

8. $21|g| + 7 = 49$

9. $-9 + 6|x| = 15$

10. $12 - 6|w| > -12$

11. $31 > 13 + 9|r|$

12. $-30 = 21 - 3|t|$

13. $9|x| - 19 < 35$

14. $-13|c| + 21 \geq -31$

15. $5 - 11|k| < -17$

16. $-42 + 14|p| = 14$

17. $15 < 3|b| + 6$

18. $9 + 5|q| = 29$

19. $-14|y| - 38 < -45$

20. $36 = 4|s| + 20$

21. $20 \leq -60 + 8|e|$

18.5   More Solving Equations and Inequalities with Absolute Values

Now, look at the following examples in which numbers and variables are added or subtracted within the absolute value symbols ($||$).

Example 9: $|3x - 5| = 10$ Remember an equation with absolute value symbols has two solutions.

Step 1: $3x - 5 = 10$ To find the first solution, remove the absolute value
$3x - 5 + 5 = 10 + 5$ symbol and solve the equation.
$\dfrac{3x}{3} = \dfrac{15}{3}$

$x = 5$

Step 2: $-(3x - 5) = 10$ To find the second solution, solve the equation for the
$-3x + 5 = 10$ negative of the expression in absolute value symbols.
$-3x + 5 - 5 = 10 - 5$
$-3x = 5$

$x = -\dfrac{5}{3}$

Solutions: $x = \left\{ 5, -\dfrac{5}{3} \right\}$

Example 10: $|5z - 10| < 20$ Remove the absolute value symbols and solve the inequality.

Step 1: $5z - 10 < 20$
$5z - 10 + 10 < 20 + 10$
$\dfrac{5z}{5} < \dfrac{30}{5}$

$z < 6$

Step 2: $-(5z - 10) < 20$ Next, solve the equation for the negative of the
$-5z + 10 < 20$ expression in the absolute value symbols.
$-5z + 10 - 10 < 20 - 10$

$\dfrac{-5z}{5} < \dfrac{10}{5}$
$-z < 2$
$z > -2$

Solution: $-2 < z < 6$

Example 11: $|4y + 7| - 5 > 18$

Step 1: $4y + 7 - 5 + 5 > 18 + 5$ Remove the absolute value symbols and solve the
$4y + 7 > 23$ inequality.
$4y + 7 - 7 > 23 - 7$
$4y > 16$
$y > 4$

Step 2: $-(4y + 7) - 5 > 18$ Solve the inequality for the negative of the
$-4y - 7 - 5 + 5 > 18 + 5$ expression in the absolute value symbols.
$-4y - 7 + 7 > 23 + 7$
$-4y > 30$
$y < -7\frac{1}{2}$

Solutions: $y > 4$ or $y < -7\frac{1}{2}$

Solve the following equations and inequalities below.

1. $-4 + |2x + 4| = 14$

2. $|4b - 7| + 3 > 12$

3. $6 + |12e + 3| < 39$

4. $-15 + |8f - 14| > 35$

5. $|-9b + 13| - 12 = 10$

6. $-25 + |7b + 11| < 35$

7. $|7w + 2| - 60 > 30$

8. $63 + |3d - 12| = 21$

9. $|-23 + 8x| - 12 > +37$

10. $|61 + 20x| + 32 > 51$

11. $|4a + 13| + 31 = 50$

12. $4 + |4k - 32| < 51$

13. $8 + |4x + 3| = 21$

14. $|28 + 7v| - 28 < 77$

15. $|62p + 31| + 43 = 136$

16. $18 - |6v + 22| < 22$

17. $12 = 4 + |42 + 10m|$

18. $53 < 18 + |12e + 31|$

19. $38 > -39 + |7j + 14|$

20. $9 = |14 + 15u| + 7$

21. $11 - |2j + 50| > 45$

22. $|35 + 6i| - 3 = 14$

23. $|26 - 8r| - 9 > 41$

24. $|25 + 6z| - 21 = 28$

25. $12 < |2t + 6| - 14$

26. $50 > |9q - 10| + 6$

27. $12 + |8v - 18| > 26$

28. $-38 + |16i - 33| = 41$

29. $|-14 + 6p| - 9 < 7$

30. $28 > |25 - 5f| - 12$

18.6   Solving Absolute Value Equations Graphically

To solve an absolute value equation graphically, consider one side of the equation as one function and the other side of the equation as another function. Then graph the two functions and see where they intersect.

Example 12:   Solve the absolute value equation $|x - 8| = 6$ graphically.

Step 1:   First, rewrite the absolute value part of the function as $f(x) = |x - 8|$. Graph the function.

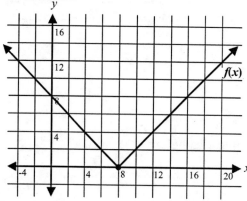

Step 2:   Now, rewrite the other side of the equation as $g(x) = 6$. Graph the function on the same graph.

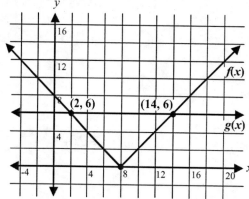

The solutions are the x-values when the y-value is 6. Looking at the graph, we can see that the two functions intersect at the points $(2, 6)$ and $(14, 6)$, so the solutions to the equation are $x = 2$ or $x = 14$.

Graph each of the following absolute value equations to find the solution, then check your answer analytically.

1. $|x + 4| = 11$

2. $|x - 3| = 8$

3. $|x - 5| = -1$

4. $|x + 10| = 2$

5. $|x - 7| = 0$

6. $|x + 6| = 13$

Example 13:   Solve the absolute value equation $-5|x + 9| + 2 = -13$ graphically.

Step 1:   First, get the absolute value part of the equation to one side of the equation by itself.
Subtract 2 from both sides of the equation so that it becomes $-5|x + 9| = -15$, and then divide both sides of the equation by -5 so that it becomes $|x + 9| = 3$.

Step 2:   Now, rewrite the absolute value part of the function as $f(x) = |x + 9|$. Graph the function.

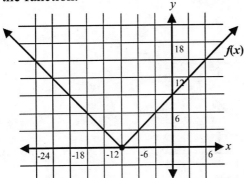

Step 3:   Now, rewrite the other side of the equation as $g(x) = 3$. Graph the function on the same graph.

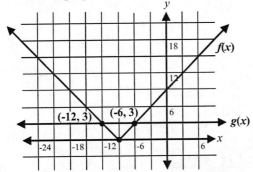

The solutions are the x-values when the y-value is 3. Looking at the graph, we can see that the two functions intersect at the points $(-12, 3)$ and $(-6, 3)$, so the solutions to the equation are $x = -12$ or $x = -6$.

Graph each of the following absolute value equations to find the solution, then check your answer analytically.

1. $6|x + 2| - 1 = 11$

2. $-3|x - 8| + 4 = -17$

3. $\frac{3}{5}|x + 10| = 24$

4. $7|x - 1| - 5 = 9$

5. $-2|x| + 17 = -19$

6. $8|x + 7| - 2 = 46$

18.7   Solving Absolute Value Inequalities Graphically

To solve an absolute value inequality graphically, consider one side of the inequality as one function and the other side of the inequality as another function. Then graph the two functions and see where they meet the conditions of the inequality.

Example 14:   Solve the absolute value inequality $|x + 5| < 2$ graphically.

Step 1:   First, rewrite the absolute value part of the function as $f(x) = |x + 5|$. Graph the function.

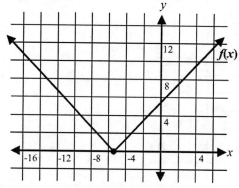

Step 2:   Now, rewrite the other side of the inequality as $g(x) = 2$. Graph the function on the same graph.

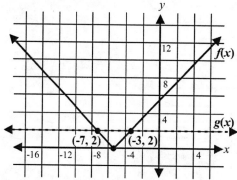

A dotted line is graphed at $g(x) = 2$. All of the x-values on the graph $f(x) = |x + 5|$ that are below (or less than) the graph of $g(x) = 2$ are solutions. The solution to the inequality is $-7 < x < -3$.

Graph each of the following absolute value inequalities to find the solution, then check your answer analytically.

1. $|x - 7| < 8$

2. $|x + 11| \leq 3$

3. $|x + 9| < 16$

4. $|x - 2| < 21$

5. $|x - 12| \leq 1$

6. $|x + 4| < 17$

Example 15:   Solve the absolute value inequality $7\,|x - 2| - 8 \geq 6$ graphically.

Step 1:   First, get the absolute value part of the inequality to one side of the equation by itself.
First add 8 to both sides of the inequality $7\,|x - 2| - 8 \geq 6$ so that it becomes $7\,|x - 2| \geq 14$, and then to divide both sides of the inequality by 7 so that it becomes $|x - 2| \geq 2$.

Step 2:   Now, rewrite the absolute value part of the function as $f(x) = |x - 2|$. Graph the function.

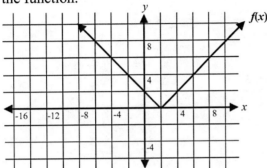

Step 3:   Now, rewrite the other side of the inequality as $g(x) = 2$. Graph the function on the same graph.

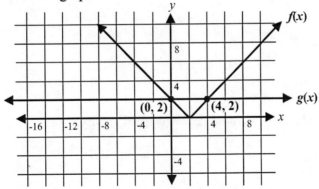

All of the x-values on the graph $f(x) = |x - 2|$ that are above or on (greater than or equal to) the graph of $g(x) = 2$ are solutions. The solution to the inequality is $x \leq 0$ or $x \geq 4$.

Graph each of the following absolute value inequalities to find the solution, then check your answer analytically.

1. $3\,|x + 9| > 24$

2. $-4\,|x - 10| + 5 \geq -11$

3. $2\,|x - 6| - 13 > 1$

4. $8\,|x + 5| + 4 \geq 36$

5. $9\,|x + 1| - 2 \geq 43$

6. $5\,|x - 18| - 6 > 9$

Chapter 18 Review

Solve the following equations and inequalities below.

1. $-11\,|k| < -22$

2. $18 - 3\,|w| > -18$

3. $2\,|2(6x + 1)| = 6$

4. $21 = -4 + |5x + 5|$

5. $\left|x + \frac{7}{10}\right| = 1$

6. $\left|\frac{7x}{2} + 7\right| = 4$

7. $|3x + 2| - 4 \geq -2$

8. $|10x + 4| < 7$

9. $|-3x - 7| \geq 11$

10. $4\,|-6x + 8| > 12$

11. $|5(3x - 1)| = 20$

12. $\left|\frac{x}{6} - 16\right| \leq 5$

13. $|6x + 1| = 11$

14. $7\,|4x + 5| \geq 35$

15. $|15x - 2| > \frac{3}{4}$

16. $|4x - 3| = 10$

17. $|-2x + 9| = 13$

18. $\left|\frac{2x}{5} - 8\right| = 9$

Solve each of the following absolute value equations graphically.

19. $|x + 16| = 2$

20. $|x - 9| = 5$

21. $4\,|x - 3| = 52$

22. $-7\,|x + 5| + 2 = -12$

23. $\frac{1}{18}\,|3(x + 5)| + 6 = 9$

24. $8\,|2(x - 1)| - 1 = 15$

Solve each of the following absolute value inequalities graphically.

25. $|x - 9| \leq 22$

26. $|x + 17| > 31$

27. $3\,|x - 7| - 19 < 5$

28. $4\,|x + 13| - 24 > 8$

29. $8\,|5(x - 7)| - 29 < 11$

30. $7\,|2(x + 8)| + 3 > 59$

Answer the following questions about the non-linear graphs.

Notre Pere College Preparatory Academy uses a 4.0 scale for its grades, to emphasize its college preparatory curriculum. Mr. Beau teaches the senior economics class, and he uses the grading scale graphed below.

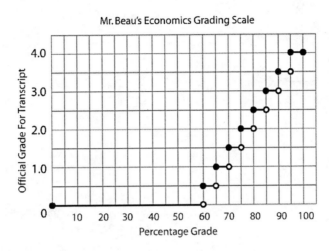

31. MaryLee scored an average of 83% in the course. What was her official 4.0 scale grade?

32. What the domain and range of this function?

Answer questions 33–34 about the graph.

The graph of function $y = \begin{cases} -x & \text{where} \quad x < 0 \\ x^2 - 4 & \text{where} \quad x \geq 0 \end{cases}$ is shown below.

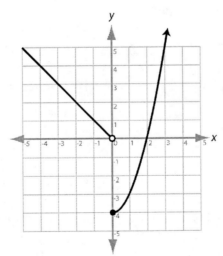

33. Where is the graph discontinuous?

34. What is (are) the zero(s) of this graph?

Chapter 18 Test

1. Which of the following absolute value equations can be solved with the graph shown?

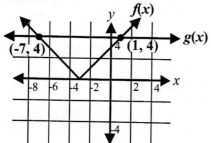

(A) $|x - 4| = 3$
(B) $|x - 3| = 4$
(C) $|x + 3| = 4$
(D) $|x + 4| = 3$

2. What is the solution to the absolute value inequality $|3(5x + 8)| < 42$?

(A) $x < -\frac{22}{5}$ or $x > \frac{6}{5}$

(B) $-\frac{22}{5} < x < \frac{6}{5}$

(C) $x < \frac{6}{5}$ or $x > \frac{22}{5}$

(D) $\frac{6}{5} < x < \frac{22}{5}$

3. If $3|x - 5| + 7 > 49$, and if $f(x) = |x - 5|$ and $g(x) = 14$, for what values of x is the graph of $f(x)$ above the graph of $g(x)$?

(A) $x < -19$ or $x > 9$
(B) $-19 < x < 9$
(C) $x < -9$ or $x > 19$
(D) $-9 < x < 19$

4. Which of the following absolute value inequalities has the solution $-14 \leq x \leq 10$?

(A) $-6|x - 2| \leq -72$
(B) $-6|x - 2| \geq -72$
(C) $-6|x + 2| \leq -72$
(D) $-6|x + 2| \geq -72$

5. The graph of the function $f(x) = |x + \frac{1}{5}|$ intersects the graph of the function $g(x) = a$ at one point. If $|x + \frac{1}{5}| \leq a$, what is the value of a?

(A) $-\frac{1}{5}$

(B) 0

(C) $\frac{1}{5}$

(D) 5

6. Solve: $39 + |10x - 8| > 41$

(A) $x = 1$
(B) $x > 1$ or $x < \frac{3}{5}$
(C) $x > \frac{3}{5}$
(D) $x > 1$ or $x < -1$

7. Solve: $|4x + 13| = 5$

(A) $x = -2$
(B) $x = 2$ or $x = 4.5$
(C) $x = -4.5$
(D) $x = -4.5$ or $x = -2$

8. Solve: $4|7x - 6| \geq 60$

(A) $x \leq -\frac{9}{7}$

(B) $x \leq -\frac{9}{7}$ or $x \geq 3$

(C) $x \leq 3$

(D) $-\frac{9}{7} \leq x \leq 3$

9. Solve: $|6(3x + 1)| = 42$

(A) $x = -\frac{8}{3}$ or $x = 2$

(B) $x = \frac{8}{3}$ or $x = -2$

(C) $x = 2$

(D) $x = -\frac{8}{3}$

Holly's family goes to the BWI airport to pick up her Grandma Jean, who is flying in today. They park in the Hourly parking garage, only to find out that Grandma Jean's flight is delayed. The graph below shows the cost of parking in the Hourly parking lot.

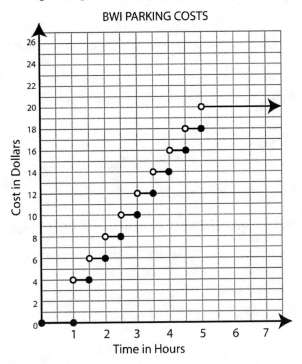

10. According to the graph, if Holly's family is parked for a total of 3 hours and 15 minutes, how much will they pay for parking?

   (A) $12
   (B) $13
   (C) $14
   (D) $16

11. What is the maximum amount Holly's family could pay for parking today in the Hourly parking lot?

   (A) $0
   (B) $20
   (C) $80
   (D) $108

Chapter 19
Using Algebra for Data Analysis

This chapter covers the following Georgia Performance Standards:

MA1D	Data Analysis and Probability	MA1D5a
		MA1D5b
		MA1D5c

19.1   Writing an Equation From Data

Bivariate data are data that involve two variables that may be related to each other. Data is often written in a two-column format. If the increases or decreases in the ordered pairs are at a constant rate, then a linear equation for the data can be found.

Example 1:   Write an equation for the following set of data.

Dan set his car on cruise control and noted the distance he went every 5 minutes.

Minutes in operation (x)	Odometer reading (y)
5	28, 490
10	28, 494

Step 1:   Write two ordered pairs in the form (minutes, distance) for Dan's driving, $(5, 28490)$ and $(10, 28494)$, and find the slope.

$$m = \frac{28494 - 28490}{10 - 5} = \frac{4}{5}$$

Step 2:   Use the ordered pairs to write the equation in the form $y = mx + b$. Place the slope, m, that you found and one of the pairs of points as x_1 and y_1 in the following formula, $y - y_1 = m(x - x_1)$.

$y - 28490 = \frac{4}{5}(x - 5)$

$y - 28490 = \frac{4}{5}x - 4$

$y - 28490 + 28490 = \frac{4}{5}x - 4 + 28490$

$y + 0 = \frac{4}{5}x + 28486$

$y = \frac{4}{5}x + 28486$

Write an equation for each of the following sets of data, assuming the relationship is linear.

1. **Doug's Doughnut Shop**

Years in Business	Total Sales
1	$55,000
4	$85,000

6. **Stepping on the Accelerator**

Seconds	MPH
4	35
7	62

2. **Gwen's Green Beans**

Days Growing	Height in Inches
2	5
6	12

7. **Aristotle's Closet**

Shirts	Price
3	$13.25
8	$32.00

3. **At the Gas Pump**

Gallons Purchased	Total Cost
5	$18.25
7	$25.55

8. **Rodriguez Family Vacation**

Hours Driven	Odometer Reading
1	263
3	423

4. **Jim's Depreciation on his Jet Skis**

Years	Value
1	$4,500
6	$2,500

9. **DJ's Dairy**

Gallons of Milk	Price
2	$7.58
4	$14.16

5. **Stepping on the Brakes**

Seconds	MPH
2	51
5	18

10. **Wall-to-Floor Decor**

Months in Business	Total Sales
4	$46,000
9	$84,000

19.2   Graphing Linear Data

Many types of data are related by a constant ratio. As you learned on the previous page, this type of data is linear. The slope of the line described by linear data is the ratio between the data. Plotting linear data with a constant ratio can be helpful in finding additional values.

Example 2:   A department store prices socks per pair. Each pair of socks costs $0.75. Plot pairs of socks versus price on a Cartesian plane.

Step 1:   Since the price of the socks is constant, you know that one pair of socks costs $0.75, 2 pairs of socks cost $1.50, 3 pairs of socks cost $2.25, and so on. Make a list of a few points.

Pair(s) x	Price y
1	0.75
2	1.50
3	2.25

Step 2:   Plot these points on a Cartesian plane, and draw a ray through the points.

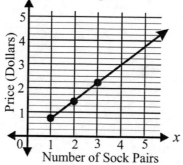

Example 3:   What is the slope of the data in the example above? What does the slope describe?

Solution: You can determine the slope either by the graph or by the data points. For this data, the slope is 0.75. Remember, slope is rise/run. For every $0.75 going up the y-axis, you go across one pair of socks on the x-axis. The slope describes the price per pair of socks.

Example 4:   Use the graph created in the above example to answer the following questions. How much would 5 pairs of socks cost? How many pairs of socks could you purchase for $3.00? Extending the line gives useful information about the price of additional pairs of socks.

Solution 1:   The line that represents 5 pairs of socks intersects the data line at $3.75 on the y-axis. Therefore, 5 pairs of socks would cost $3.75.

Solution 2:   The line representing the value of $3.00 on the y-axis intersects the data line at 4 on the x-axis. Therefore, $3.00 will buy exactly 4 pairs of socks.

Use the information given to make a line graph for each set of data, and answer the questions related to each graph.

1. The diameter of a circle compared with the circumference of a circle is a constant ratio. Use the data given below to graph a line to fit the data. Extend the line, and use the graph to answer the next question.

Circle

Diameter	Circumference
4	12.56
5	15.70

2. Using the graph of the data in question 1, estimate the circumference of a circle that has a diameter of 3 inches.

3. If the circumference of a circle is 3 inches, about how long is the diameter?

4. What is the slope of the line you graphed in question 1?

5. What does the slope of the line in question 4 describe?

6. The length of a side on a square and the perimeter of a square are constant ratios to each other. Use the data below to graph this relationship.

Square

Length of side	Perimeter
2	8
3	12

7. Using the graph from question 6, what is the perimeter of a square with a side that measure 4 inches?

8. What is the slope of the line graphed in question 6?

9. Conversions are often constant ratios. For example, converting from pounds to ounces follows a constant ratio. Use the data below to graph a line that can be used to convert pounds to ounces.

Measurement Conversion

Pounds	Ounces
2	32
4	64

10. Use the graph from question 9 to convert 40 ounces to pounds.

11. What does the slope of the line graphs for question 9 represent?

12. Graph the data below, and create a line that shows the conversion from weeks to days.

Time

Weeks	Days
1	7
2	14

13. About how many days are in $2\frac{1}{2}$ weeks?

19.3   Modeling Data with Linear Functions

When the dependent variable (y) changes at a constant rate in relation to the change in an independent variable (x), the relationship can be modeled with a linear function.

Example 5:   A gas-station owner has noticed that as the price of a gallon of gas has increased, the number of gallons of gas purchased at his station has decreased. The table below shows the average price of a gallon of gas at the station for the last 10 weeks and the number of gallons of gas purchased at the station during that week.

	Average Price	Gallons Purchased
Week 1	$2.40	3,200
Week 2	$2.51	3,145
Week 3	$2.69	3,055
Week 4	$3.00	2,900
Week 5	$3.14	2,830
Week 6	$3.36	2,720
Week 7	$3.59	2,605
Week 8	$3.71	2,545
Week 9	$4.01	2,395
Week 10	$4.20	2,300

Can the relationship between the average weekly price of a gallon of gas at the station and the number of gallons purchased for the last 10 weeks be modeled with a linear function? If so, plot the data and draw the line.

Step 1:   Determine if the relationship can be modeled with a linear function.

From the table, we see that for every $0.01 increase in the average weekly price of a gallon of gas, there has been a 5-gallon decrease in the number of gallons purchased per week. Since the dependent variable (the number of gallons purchased) decreases at a constant rate in relation to an increase in the independent variable (the average price of a gallon of gas), this is an inverse relationship that can be modeled with a linear function.

Step 2:   Plot the data and draw the line.
The average price of a gallon of gas is the independent variable (x), and the number of gallons purchased is the dependent variable y.

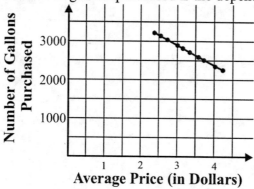

Example 6:   A new barbershop has been open for 5 days, and the number of daily patrons to the barbershop has been increasing at a constant rate each day. The relationship between the number of days the barbershop has been open and the number of daily patrons can be modeled with a linear function as shown below. What is the equation of the linear function?

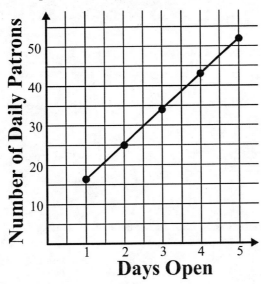

Step 1:   Determine the slope of the equation of the line.

Looking at the graph, we can find two points. Two points on the line are $(1, 16)$ and $(3, 34)$. Calculate the slope.

$$\frac{\text{change in } y}{\text{change in } x} = \frac{34 - 16}{3 - 1} = 9.$$

So far, the equation of the line is $y = 9x + b$, where b is the y-intercept.

Step 2:   Determine the y-intercept of the equation of the line.

The equation of the line is $y = 9x + b$, and one of the points on the line is $(1, 16)$. Using this information, substitute the point into the equation to solve for b.

$16 = 9(1) + b$

$16 = 9 + b$

$16 - 9 = b$

$b = 7$

The equation of the line is $y = 9x + 7$.

**Note: Even though the y-intercept of the equation of the line is 7, the line segment that is the model for this situation does not actually cross the y-axis, since the first day the barbershop had any patrons was day 1.

Determine whether or not each of the following situations can be modeled with a linear function.

1. For every $0.50 increase in the price of a loaf of bread, the number of loaves purchased per day is cut in half.

2. Every 5 minutes the number of cookies in a cookie jar decreases by 3.

3. Every time the number of fans in a stadium increases by 1,000, the noise level increases by 10 percent.

4. For every mile a runner runs, she burns 100 calories.

5. For every new employee a company hires, it spends $4,000 on training.

6. A car is travelling at 25 miles per hour.

7. The area of a square is equal to the length of a side of the square multiplied by itself.

8. The total number of visits to a website is quadrupling every month.

9. To produce a ton of paper, 24 trees must be cut down.

Find the equation of the linear function that can be used to model each situation.

10. (1 hr, 5 meters), (4 hrs, 41 meters)

11. ($1.50, 78 gallons), ($1.90, 54 gallons)

12. (26 people, 6 cans), (38 people, 9 cans)

13. (7 boxes, 44 lbs), (12 boxes, 79 lbs)

14. (88°, 70 customers), (95°, 56 customers)

15. (3 days, 120 clicks), (9 days, 200 clicks)

19.4   Making Scatter Plots

A scatter plot is a graph of the relationship between two variables.

Example 7:   Below is the height and weight of 14 people. Use this data to make a scatter plot.

height (in)	36	54	60	66	39	48	44	72	75	62	61	45	50	59
weight (lb)	40	104	107	150	48	77	62	195	205	115	112	65	85	106

Step 1:   To make a scatter plot, make a Cartesian plane and label the vertices using the two variables given in the table above.
Then plot the points on the grid.

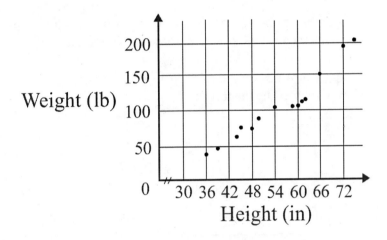

Step 2:   Determine the relationship between the two variables by looking at the graph. We can tell that as the height increases, weight increases.

Draw a scatter plot for the given data.

1.

Age	6	11	19	5	8	12	14	7	13	17	9
Height	46	56	64	42	52	56	63	51	60	63	54

2.

Gas Price	2.50	3.60	2.63	3.00	2.20	2.89	2.36	3.41	3.26
Dist. Driven/Week	250	75	225	175	300	200	280	125	150

3.

Height	62	70	76	65	74	75	73	71	77	66	72	63
Pts. Scored	21	10	15	30	40	6	11	22	14	17	18	26

19.5   Interpreting Data in Scatter Plots

You have already learned that scatter plots show the relationship between two variables. Now, you will learn to explain the relationship between variables.

Example 8:   The graph below shows the relationship between height and age.

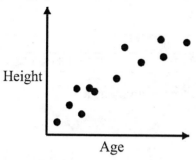

Although it isn't linear, there is clearly a positive relationship between age and height. This means that as age increases, height increases.

Example 9:   The graph below shows the relationship between price of an object and the number purchased by customers.

This illustrates a negative relationship. This means as price of an object increases, the number purchased decreases. In other words, if the price of an object goes up, fewer people will buy that object.

Example 10:   The graph below shows the relationship between number of points scored on a test and height

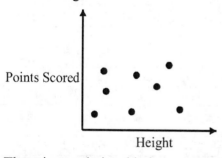

There is no relationship between height and points scored because there is no definable pattern. Therefore, height and number of points scored on a test are not correlated.

Explain the relationship between the variables in each of the following scatter plots.

1.

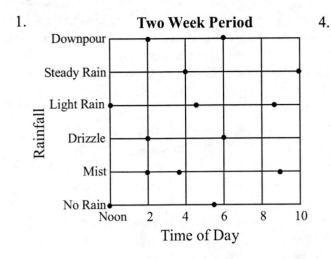

2.

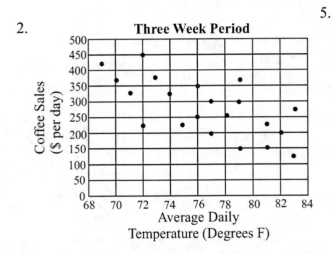

3.

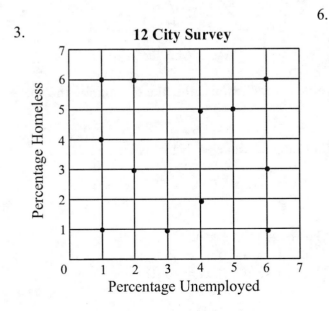

4.

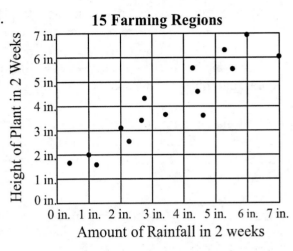

5.

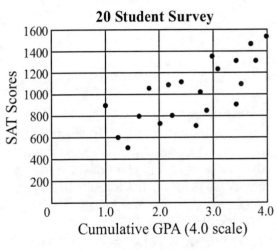

6.

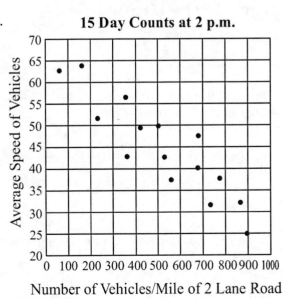

19.6   The Line of Best Fit

At this point, you now understand how to plot points on a Cartesian plane. You also understand how to find the data trend on a Cartesian plane. These skills are necessary to accomplish the next task, determining the line of best fit. The line of best fit is a straight line that demonstrates the relationship between two variables. The line does not necessarily divide the plotted points into two areas, but this sometimes is the best way to estimate the line of best fit.

To estimate the line of best fit, you must first draw a scatter plot of all data points. Once this is accomplished, draw an oval around all of the points plotted. Draw a line through the points in such a way that the line separates half the points from one another. You may now use this line to answer questions.

Example 11:   The following data set contains the heights of children between 5 and 13 years old. Make a scatter plot and draw the line of best fit to represent the trend. Using the graph, determine the height for a 14-year old child.

Age 5: 4'6",4'4",4'5"   Age 8: 4'8", 4'6", 4'7"   Age 11: 5'0", 4'10"
Age 6: 4'7", 4'5", 4'6"   Age 9: 4'9", 4'7", 4'10"   Age 12: 5'1", 4'11", 5'0", 5'3"
Age 7: 4'9", 4'7", 4'6", 4'8" Age 10: 4'9", 4'8", 4' 10"   Age 13: 5'3", 5'2", 5'0", 5'1"

In this example, the data points lay in a positive sloping direction. To determine the line of best fit, all data points were circled, then a line of best fit was drawn. Half of the points lay below, half above the line of best fit drawn bisecting the narrow length of the oval. The is called "eye-balling."
To find the height of a 14–year old, simply continue the line of best fit forward. In this case, the height is 62 inches.

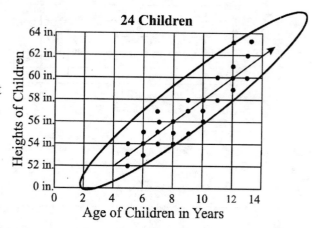

Plot the data sets below, then draw the line of best fit. Next, use the line to estimate the value of the next measurement.

1. Selected values of the Sleekster Brand Light Compact Vehicles: New Vehicle: $13,000.
   1 year old: $12,000, $11,000, $12,500   3 year old: $8,500, $8,000, $9,000
   2 year old: $9,000, $10,500, $9,500   4 year old: $7,500, $6,500, $6,000
   5 year old: ?

2. The relationship between string length and kite height for the following kites:
   (L = 500 ft, H = 400 ft) (L = 250 ft, H = 150ft) (L = 100 ft, H = 75 ft)
   (L = 500ft, H = 350 ft) (L = 250 ft, H = 200 ft) (L = 100 ft, H = 50 ft)
   (L = 600 ft, H = ?)

19.7   More Lines of Best Fit

Relationships that can be modeled with linear functions usually are not exactly linear. In other words, the linear model is only an approximation, and there are points that do not lie exactly on the line. When this is the case, methods such as eyeballing (which we studied in the previous section) and finding the median-median line can be used to determine the equation of the linear model.

Example 12:   Jake rode his bicycle for a total of 240 minutes, and he travelled a total of 40 miles. However, he did not travel at a constant speed, so the graph representing his distance travelled as a function of time is not exactly linear. A scatter plot representing Jake's distance travelled in miles as a function of time passed in minutes is shown.

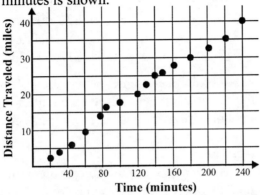

Use eyeballing to determine the equation of the linear function that can be used to model the relationship between the time passed and the distance travelled.

Step 1:   Determine the equation of a line that would be a good approximate representation of the relationship being modeled.
By looking at the points included in the scatter plot, it appears that if the line $y = \frac{1}{6}x$ were drawn, about the same number of points would be above the line as would be below the line. For this reason, the line $y = \frac{1}{6}x$ would be a good representation of the relationship between the time passed and the distance travelled.

Step 2:   Draw the line on the graph.

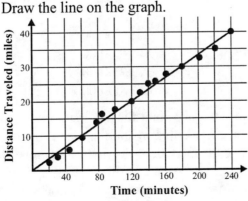

Since about the same number of points are, in fact, above the line as are below the line, the line $y = \frac{1}{6}x$ is an appropriate model.

Example 13:   Over the course of the last 9 months, a retail store has been increasing the amount of money it spends on advertising, and it has noticed that for the most part, the more money it spends, the more revenue is generated by its store. This is shown in the table below.

	Amount Spent on Advertising	Store Revenue
Month 1	$3,000.00	$30,000.00
Month 2	$4,500.00	$59,000.00
Month 3	$5,500.00	$55,000.00
Month 4	$7,000.00	$62,000.00
Month 5	$7,500.00	$31,000.00
Month 6	$9,500.00	$68,000.00
Month 7	$10,000.00	$75,000.00
Month 8	$11,500.00	$71,000.00
Month 9	$12,000.00	$77,000.00

Find the **median-median line** to determine the equation of the linear model that best represents the relationship between the amount of money spent on advertising and the amount of revenue generated by the store.

Step 1:   Divide the data into 3 groups.

Since there are 9 columns, the first group is the first 3 columns, the second group is the next 3 columns, and the third group is the last 3 columns.

Step 2:   Determine the summary point (median) for each group.

To determine the summary point for a group, the table should be written so that the x-coordinates of the data points are in order from least to greatest (this is already done).

Amount Spent on Advertising	Store Revenue
$3,000.00	$30,000.00
$4,500.00	$59,000.00
$5,500.00	$55,000.00
$7,000.00	$62,000.00
$7,500.00	$31,000.00
$9,500.00	$68,000.00
$10,000.00	$75,000.00
$11,500.00	$71,000.00
$12,000.00	$77,000.00

The x-coordinate of the summary point for the first group is the median of the first 3 x-coordinates in the first column, and the y-coordinate of the summary point for the first group is the median of the first three y-coordinates in the last column. Therefore, the summary point for the first group is ($4,500, $55,000). The summary points for the next two groups can be found in the same way, and are ($7,500, $62,000) and ($11,500, $75,000), respectively.

Step 3: Plot the three summary points, then draw a line through the first and last summary points.

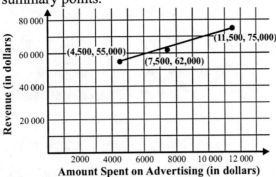

Step 4: Move the line one-third of the way toward the middle summary point to find the median-median line. The distance from all points below the line to the line should equal the distance from all points above the line to the line.

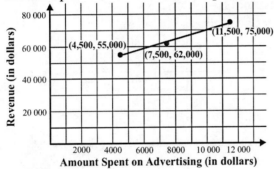

Step 5: Find the equation of the median-median line.
The slope of the median-median line equals the slope of the line that passes through the first and last summary points.

$$m = \frac{75000 - 55000}{11500 - 4500} = \frac{20000}{7000} \approx 2.86$$

Therefore, the equation of the median-median line is $y = 2.86x + b$.
The easiest way to find the y-intercept of the line is to extend the line all the way to the y-axis. The y-intercept is approximately $41,619$.
The equation of the linear model is $y = 2.86x + 41619$.
The line is graphed below along with all of the original data points.

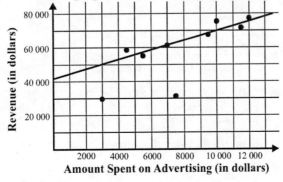

For each of the following groups of points, use eyeballing to determine the equation of the linear function that can be used to model the relationship in question.

1. ($1.50, $4.90), ($2.00, $6.00), ($2.50, $7.10), ($3.00, $8.10), ($3.50, $8.90)

2. (11.9 hr, 50 in), (13 hr, 56 in), (14.1 hr, 64 in), (15 hr, 72 in), (16 hr, 78 in)

3. (1 day, 8.2 km), (5 days, 9.8 km), (7 days, 11.3 km), (9 days, 11.7 km)

4. (6.3 gallons, 7 lbs), (8.7 gallons, 8 lbs), (12 gallons, 9.4 lbs), (15 gallons, 9.6 lbs)

5. (44 watts, $3.06), (52 watts, $3.20), (64 watts, $3.34), (76 watts, $3.60)

6. (5 yr, 22 mm), (6 yr, 26 mm), (7 yr, 33 mm), (8.3 yr, 37 mm), (8.7 yr, 42 mm)

7. (3 cars, 12.7 quarts), (4 cars, 20 quarts), (8 cars, 28 quarts), (9 cars, 35.3 quarts)

8. (2.1 wk, 1 cm), (3 wk, 2.1 cm), (4 wk, 2.9 cm), (5 wk, 4 cm), (5.9 wk, 5 cm)

9. ($2.75, 9 units), ($2.86, 19 units), ($2.95, 26 units), ($3.04, 42 units), ($3.15, 49 units)

Use the table below to answer questions 10–14.

Hours Studying	Points on Test
0.5	40
1.5	44
2	46
3	47
3.5	49
4	45
5	51
6	50
6.5	53
7.5	59
8	88
8.5	62
9	67
10	68
10.5	70
11	73
11.5	55
12	76
13	80
14	89
14.5	92

10. Determine the summary point for each of the three groups into which the data is divided.

11. Find the slope of the line that passes through the first and last summary points.

12. Find the slope of the median-median line.

13. Find the approximate y-intercept of the median-median line.

14. Find the equation of the linear model that best represents the situation.

19.8   Linear Regression

In the previous section, you saw how to estimate the line of best fit using a graph. Using a graphing calculator, you can calculate an exact line of best fit. The graphic calculator may refer to this line of best fit as "the line of regression" or "LinReg."

To find the line of best fit, you must first enter your data points into the calculator. Find your statistics (STAT) key. Pressing STAT will give you a menu where you need to find the option that allows you to enter data. On a TI-83, this is "Edit". Pressing ENTER will give you an opportunity to type in your data points in list form. Make sure that you put your x-axis variable in the first list (L1) and your y-axis variable in the second list (L2), and that they match up. For example, if your first point is $(3, 4)$, then 3 should be the first number in L1 and 4 should be the first number in L2.

Once you have entered all your data points, go back to the STAT menu and find where you can calculate a line of best fit or "linear regression." On a TI-83, press "right" and you will be under the "CALC" heading at the top of the STAT menu. Find option 4, "LinReg(ax+b)," and press ENTER. Now you will be back on the main screen with "LinReg(ax+b)" displayed. Type in the following on the TI-83: L1,L2.

Example 14:   The following data set contains the length of car trip that the Doran family has made and the number of stops that they made on each trip.

Length of trip	Number of Stops	Length of Trip	Number of Stops
2 hours	0 stops	8 hours	3 stops
5 hours	2 stops	5 hours	3 stops
4 hours	2 stops	11 hours	6 stops
10 hours	6 stops	9 hours	4 stops

Enter the data: (STAT, Edit...) Put the length of trip in L1 column and number of stops in the L2 column like the following:

L1 (x)	L2 (y)
2	0
5	2
4	2
10	6
8	3
5	3
11	6
9	4

Now calculate the line of best fit by pressing STAT and going to the CALC menu. Find LinReg(ax+b) and press enter. Now you will be back on the main screen with "LinReg(ax+b)" displayed. Type in the following on the TI-83: L1,L2.
Output looks like:
LinReg;
$y = ax + b$;
$a = 0.6083916084$;
$b = -0.8566433566$

So, the line of best fit is Stops $= 0.608 \times$ Length $- 0.857$ or $y = 0.608x - 0.857$
The slope means that for every hour longer the trip is, on average, they will stop 0.608 more times.

Let's consider how well the line predicts the data points at Length $= 5$ hours and Length $= 10$ hours.
$$y = 0.608 \times 5 - 0.857 = 2.183 \qquad y = 0.608 \times 10 - 0.857 = 5.223$$

The prediction for 5 hours was 2.183 stops, while the real observed values were 2 and 3, and the prediction for 10 hours was 5.223 stops, while the real observed value was 6. These are all quite close and show that this is a reasonable line of best fit.

Calculate the line of best fit for each situation.

1. Tabitha has to keep track of the number of minutes she reads each day and how far she gets into her book. Her observations from her most recent week of reading are graphed below.

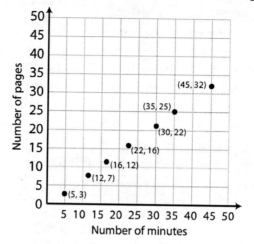

About how much of a page can she read per minute? Use the exact line of best fit to determine your answer.

2. Dr. Silverstein charted 8 of his patients' candy consumption versus their number of cavities in a year.

Patient	Candies Per Day	Cavities Per Year	Patient	Candies Per Day	Cavities Per Year
Kurt	1	1	Chris	2	0
Erin	4	2	Olivia	7	3
Sarah	4	3	Rimas	5	2
Brian	9	5	Tonya	10	4

How many cavities should Jessica have if she ate 8 candies per day? Use the exact the line of best fit to determine your answer.

3. A group of 9 students in Mrs. Van Wyck's math class were given the assignment to determine if there was a strong relationship between the number of people in their household and the amount that their household spent on groceries (excluding pet food).
Their data points were as follows (Number of people in household, cost of groceries):
(3, $143), (4, $156), (2, $89), (2, $127), (6, $201), (5, $180), (3, $171), (3, $152), (3, $135)
Mrs. Van Wyck has a household of 7 people. How many dollars would you estimate her grocery bill to be? Use the line of best fit to determine your answer.

For questions 4–8, use the linear regression feature on a graphing calculator to determine the equation of the linear function that can be used to model the relationship in each question.

4. (10 years, 3.5 mm), (11 years, 9.2 mm), (13 years, 14.5 mm), (14 years, 22.8 mm)

5. (2.2 min, 75 beats), (4 min, 149 beats), (6 min, 213 beats), (8.8 min, 298 beats)

6. ($2.47, 11 pts), ($2.55, 16 pts), ($2.58, 19 pts), ($2.61, 20 pts), ($2.66, 27 pts)

7. (4 laps, 42 calories), (10 laps, 113 calories), (19 laps, 176 calories), (23 laps, 258 calories)

8. (2 carries, 9 yards), (3 carries, 21 yards), (7 carries, 28 yards), (8 carries, 43 yards)

19.9   Modeling Data with Quadratic Functions

When the change in a dependent variable changes at a constant rate in relation to the change in an independent variable, the relationship can be modeled with a quadratic function.

Example 15:   A backyard swimming pool originally containing 128 m^3 of water developed a small leak, so the amount of water in the swimming pool decreased with time. The table below shows the amount of water that was in the swimming pool after certain numbers of hours had passed since the leak began.

Hours Since Leak Began	Amount of Water in Pool in m^3
0	128
1	126
2	120
3	110
4	96
5	78
6	56
7	30
8	0

Can the relationship between the number of hours after the leak began and the amount of water in the pool be modeled with a quadratic function? If so, plot the data and draw the parabola.

Step 1:   Determine if the relationship can be modeled with a quadratic function.
From hour 1 to hour 2, the change in the amount of water in the pool was -6 m^3, from hour 2 to hour 3 it was -10 m^3, from hour 3 to hour 4 it was -14 m^3, and so on. The change in the amount of water in the pool is changing at a rate of -4 m^3 per hour. Since the change in the dependent variable decreases at a constant rate in relation to an increase in the independent variable, the relationship can be modeled with a quadratic function.

Step 2:   Plot the data and draw the line.
The number of hours after the leak began is the independent variable (x), and the amount of water in the pool is the dependent variable (y).

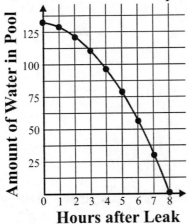

Example 16:   A blog has been online for 5 days, and the change in the number of comments posted to the blog from one day to the next has been increasing at a constant rate. The relationship between the number of days the blog has been online and the total number of comments posted can be modeled with a quadratic function as shown below. What is the equation of the quadratic function?

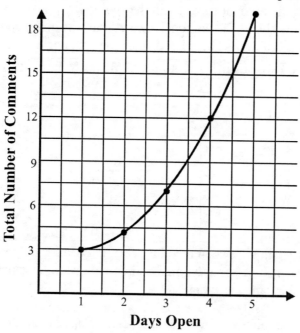

Days Open

Step 1:   Find the vertex of the parabola.

Looking at the graph, we see the right side of a parabola with a vertex of $(1, 3)$. If this is the case, the equation of the parabola is $y = a(x - 1)^2 + 3$, where a is a constant.

Step 2:   Substitute one point into the equation of the parabola and solve for a.

One point on the parabola is $(4, 12)$. Substitute and solve for a.

$12 = a(4 - 1)^2 + 3$

$12 = a(3)^2 + 3$

$12 = 9a + 3$

$12 - 3 = 9a + 3 - 3$

$9 = 9a$

$a = 1$

Since $a = 1$, the equation of the parabola is $y = 1(x - 1)^2 + 3$, or $y = (x - 1)^2 + 3$.
If all the other points on the parabola are tested, they all satisfy this equation, so the equation is, in fact, correct.

Determine whether or not each of the following situations can be modeled with a quadratic function.

1. Every month the number of checks in a check book decreases by 4.

2. For every $50 decrease in the price of a computer, the change in the number of computers sold increases by 12.

3. The total number of donations made to a charitable organization is increasing at a rate of 28 per day.

4. A person's heart is beating at a rate of 70 beats per minute.

5. For every hour that passes, the change in temperature decreases by 0.2 degrees.

6. For every week that passes, the total amount of rainfall increases by 0.1 inches.

7. Every day the change in the number of visitors to a park over the previous day increases by 18.

8. For every student a college accepts, it spends $50 on facility maintenance.

9. For every point won in a debate, the change in a politician's approval rating increases by 0.5 percent.

What is the equation of the quadratic function that can be used to model the situation with each of the following pairs of data points? Assume the first data point is the vertex of the parabola.

10. (3 years, 5 km), (8 years, 55 km)

11. (1 mile, 6 min), (9 miles, 70 min)

12. (3 signs, 10 people), (4 signs, 9 people)

13. ($3.90, 44 units), ($4.30, 60 units)

14. (2 games, 7 points), (6 games, 55 points)

15. (5 cases, $70.00), (7 cases, $58.00)

19.10   Curve of Best Fit

Sometimes, the trend of best fit is not best described by a line, but by some other kind of curve. Some questions will ask you to read this curve on a graph, predict missing values, and to interpret the meaning in real-life terms.

Example 17:   The great quarterback "Touchdown" Tofanelli played professional football for 12 years.
The following graph shows his touchdown passes each year in the league.

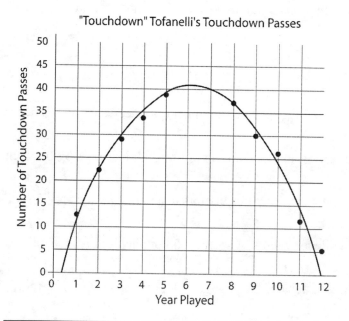

"Touchdown" Tofanelli's Touchdown Passes

Year	1	2	3	4	5	6	7	8	9	10	11	12
Number of TD passes	12	22	29	33	39	?	?	37	30	26	11	3

Step 1:   Describe the curve of best fit, explaining in what periods (if any) there was a rapid increase, a rapid decrease, a slow increase, a slow decrease, or a steady level.

The curve of best fit is shaped like a parabola, with zeroes around Years 0 and 12. There is a rapid increase in TD passes per year from Years 1 to 3, a slow increase in TD passes per year from Years 4 to 6, a slow decrease in TD passes per year from Years 7 to 9, and a rapid decrease in TD passes from Years 10 to 12. In everyday terms, Tofanelli got better every year until he hit his "prime" around Years 3, played his best years in Years 6 and 7, and lost his effectiveness after Year 10.

Step 2:   According to the curve of best fit, how many touchdown passes should he have thrown in years 6 & 7?

The curve passes at Year = 6 passes through 41 TD passes or so, and Year = 7 clearly passes through 40 TD passes.

Answer the questions by looking at the curve of best fit graphs.

1.

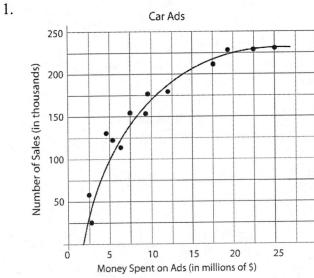

(A) A car executive wanted to see how much increasing his ad budget would affect sales. He plotted previous ad campaigns for similar cars against the number of sales. According to the curve of best fit, how much would their sales increase if they increased their ad budget from $10 million to $15 million?

(B) Explain how the amount of sales changes in comparison to the amount spent on advertising. Include an estimate of when sales level off.

2.

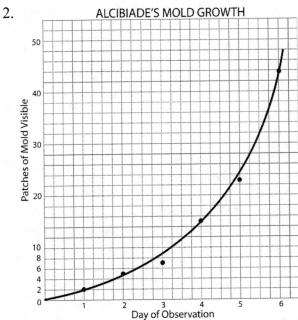

Alcibiade, the scientist, is testing the growth of a particular kind of mold. According to the curve of best fit, how many patches of mold should there be after 6 days?

19.11   Quadratic Regression

When a relationship is not exactly quadratic, quadratic regression can be used to determine the equation of the quadratic model.

Example 18:   A college has offered a course on computer graphics for the last 5 semesters, and the enrollment in the course has increased with each semester. The college has found that as the enrollment in the course has increased, the number of textbooks on computer graphics sold in its bookstore has also increased. The table below shows the enrollment in the course for each of the last 5 semesters and the number of textbooks on computer graphics sold in its bookstore during that semester.

	Students Enrolled in Course	Textbooks Sold
Semester 1	9	16
Semester 2	11	19
Semester 3	16	44
Semester 4	19	71
Semester 5	25	152

Use the quadratic regression feature on a graphing calculator to determine the equation of the quadratic function that can be used to model the relationship between the number of students enrolled in the computer graphics course and the number of textbooks sold.

Enter the data: (STAT, Edit...) Put the length of trip in L1 column and number of stops in the L2 column like the following:

L1 (x)	L2 (y)
9	16
11	19
16	44
19	71
25	152

Now calculate the curve of best fit by pressing STAT and going to the CALC menu. Find QuadReg and press enter. Now you will be back on the main screen with "QuadReg" displayed. Type in the following on the TI-83: L1,L2. Output looks like:

QuadReg;
$y = ax^2 + bx + c$;
$a = 0.5$;
$b = -8.5$;
$c = 52$

Therefore, the equation for this situation is $y = 0.5x^2 - 8.5x + 52$.

For each of the following groups of points, use the quadratic regression feature on a graphing calculator to determine the equation of the quadratic function that can be used to model the relationship in question.

1. (1 hr, 3 liters), (2 hrs, 7 liters), (3.5 hrs, 27 liters), (4 hrs, 34 liters), (5 hrs, 51 liters)

2. (1 day, 4 views), (3 days, 25 views), (4 days, 49 views), (5 days, 79 views), (6 days, 99 views)

3. (2 people, $42), (6 people, $250), (8 people, $670), (12 people, $1250), (15 people, $1710)

4. (2 min, 99 pgs), (4 min, 197 pgs), (7 min, 354 pgs), (8 min, 484 pgs), (9 min, 649 pgs)

19.12   Misusing Lines and Curves of Best Fit

While the lines and curves of best fit are very helpful, you must be careful not to predict values too far outside the range of data.

Example 19:   Muddy Creek High School opened in 1997. In 1997, it taught only freshmen. In 1998, it taught freshmen and sophomores. Below is a graph of the number of students in each year.

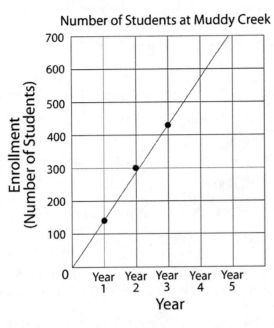

Number of Students at Muddy Creek

Year	1997	1998	1999	2000
Number of Students	140	302	435	?

About how many students should there be in the school in 2000 (year 4). Based on the line of best fit, how many students would we expect to be in the school by the year 2025 (year 29)? Is this number reasonable?

Step 1:   The exact line of best fit is $y = 147.5x - 2.67$. Whether or not this exact line is calculated, you can estimate that 140 or 145 are being added each year. The exact answer for the year 2000 is $y = 147.5 \times 4 - 2.67 = 587.33$, about 587 students. On the graph, you can see that it is between 580 and 600.

Step 2:   The same equation yields, for 2025, $y = 147.5 \times 29 - 2.67 = 4274.8$, roughly 4275 students. This number is not reasonable because the fast growth of the high school is a result of the introduction of an entirely new class each year. We have no way of knowing if and how much the high school will grow after the first four years.

Chapter 19 Review

Determine whether or not each of the following situations can be modeled with a linear function.

1. For every page a writer writes, he gets paid $10.50.

2. The number of supporters of a candidate is increasing by 5 percent every month.

3. The perimeter of a square is equal to the length of a side of the square multiplied by 4.

4. The total number of signatures on a petition is tripling every week.

Find the equation of the linear function that can be used to model a situation with the following data points.

5. (5 yrs, 277 km), (9 yrs, 437 km)

6. (34 crates, $8.50), (41 crates, $7.66)

Determine whether or not each of the following situations can be modeled with a quadratic function.

7. Every minute the change in the amount of water in a bathtub decreases by 0.5 liters.

8. A company's debt increases by $500,000 every year.

9. The change in the number of people donating blood is increasing by 14 every quarter.

10. The number of mosquitoes is decreasing by 5 every minute.

Find the equation of the quadratic function that can be used to model a situation with the following data points. Assume the first data point is the vertex of the parabola.

11. (3 matches, 2 pts), (8 matches, 52 pts)

12. ($2.40, 12 bushels), ($6.90, 93 bushels)

Use the data below for questions 13–16.

(2 pages, 44 points), (9 pages, 48 points), (13 pages, 77 points), (16 pages, 53 points), (28 pages, 64 points), (41 pages, 51 points), (57 pages, 70 points), (58 pages, 73 points), (107 pages, 89 points)

13. Determine the summary point for each of the three groups into which the data is divided.

14. Find the slope of the median-median line.

15. Find the approximate y-intercept of the median-median line.

16. Find the equation of the linear model that best represents the situation.

For each of the following groups of points, use the linear regression feature on a graphing calculator to determine the equation of the linear function that can be used to model the relationship in question.

17. (13.2 days, 32 km), (14.1 days, 37 km), (14.9 days, 47 km), (16.3 days, 56 km)

18. ($62.13, 42 people), ($64.03, 48 people), ($65.98, 50 people), ($68.08, 54 people)

19. (15 games, 30 errors), (25 games, 48 errors), (32 games, 60 errors), (40 games, 83 errors)

20. (2 min, 58 gallons), (11 min, 79 gallons), (23 min, 91 gallons), (31 min, 125 gallons)

For each of the following groups of points, use the quadratic regression feature on a graphing calculator to determine the equation of the quadratic function that can be used to model the relationship in question.

21. (3.3 kg, $3.06), (4.1 kg, $5.25), (5.6 kg, $8.77), (6.2 kg, $12.01), (7.8 kg, $17.76)

22. (4 hr, 2 pints), (8 hr, 13 pints), (10 hr, 35 pints), (15 hr, 78 pints), (19 hr, 141 pints)

Explain the relationship between the variables in each of the following scatter plots.

23.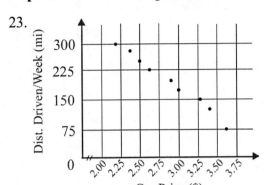

24.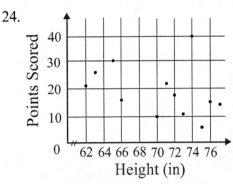

The function provided was used to model each of the following relationships. Determine the mathematical model's domain.

25. $y = 4x$; the number of DVDs watched in a collection of 92 as a function of weeks passed

26. $y = -20x^2 + 1620$; a computer's value in dollars as a function of years passed

27. $y = 3x^2$; the number of online friends made in July as a function of days passed that month

28. $y = 0.2x$; the number of hours spent dusting as a function of the number of rooms cleaned

Draw a scatter plot for the given data.

29.

Price of strawberries/lb	1.00	1.50	5.00	3.50	2.50	4.00	3.00
lbs purchased	75	65	20	40	52	30	47

30.

Height (in)	62	73	64	71	72	63	75	74	66	73	75
Homeruns hit	6	10	1	0	11	5	4	3	6	2	1

Answer the following questions about line and curve of best fit.

31. In 1870, Lonesome Dove was a prosperous Western town with 12,000 people when the railroad decided to change routes. The table below shows Lonesome Dove's population decline in subsequent decades:

Year	1870	1880	1890	1900	1910
Population of Lonesome Dove	12000	10300	8500	6900	5200

Find the exact line of best fit using a calculator, with 1870 as $x = 0$.

(A) According to the line of best fit, what was the population in 1896 (Year 26)?

(B) According to the line of best fit, what will the population be in 2010 (Year 140)? Is this number reasonable?

32. The graph below shows the median high temperature each month in Buenos Aires, Argentina.

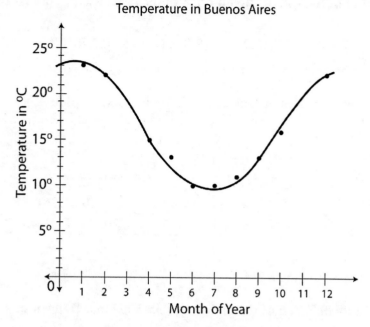

According to the curve of best fit, what should be the missing values for March and November?

Chapter 19 Test

1. Which of the following situations can best be modeled with a quadratic function?

   (A) For every month that passes, the number of songs stored on an MP3 player increases by 22.
   (B) The change in the number of students enrolled in a college over the previous semester is increasing by 58 each semester.
   (C) The amount of yogurt consumed by citizens of a country is increasing by 25,000 ounces every month.
   (D) The average number of minutes high school students sleep every night is decreasing by 3 minutes every 5 years.

2. Linear regression has been used to determine the equation of the linear function that models the situation that produces the points $(6.2$ minutes, 26 inches$)$, $(7.1$ minutes, 34 inches$)$, $(7.9$ minutes, 41 inches$)$, and $(9.3$ minutes, 50 inches$)$. What is the equation of the linear function?

   (A) $y = -21.2x + 7.7$
   (B) $y = -7.7x + 21.2$
   (C) $y = 7.7x - 21.2$
   (D) $y = 21.2x + 7.7$

3. A linear function is being used to model the number of daily visitors to a museum as a function of the admission price in dollars. If the slope of the line is -20, and if one of the points on the line is $(6, 780)$, what is the equation of the line?

   (A) $y = -20x + 120$
   (B) $y = -20x + 660$
   (C) $y = -20x + 780$
   (D) $y = -20x + 900$

4. Based on the data shown, which of the following is not a summary point used to find the median-median line that models the situation?

Times Played	Average Number of Points Scored on Video Game
1	444
2	478
4	497
6	530
7	522
8	545
10	468
13	582
17	619

   (A) (7 time played, 530 points scored)
   (B) (2 times played, 478 points scored)
   (C) (7 times played, 522 points scored)
   (D) (13 times played, 582 points scored)

5. If a change in variable x suggests a change in variable y, does the change in x cause the change in y?

   (A) Always
   (B) Sometimes
   (C) Never
   (D) Only if there is a direct relationship

6. Quadratic regression has been used to determine the equation of the quadratic function that models the situation that produces the points $(5.3$ kg, $90.55)$, $(6.2$ kg, $111.03)$, $(7.5$ kg, $151.63)$, $(8.2$ kg, $209.97)$, and $(9.6$ kg, $293.32)$. What is the equation of the quadratic function?

   (A) $y = 10.9x^2 - 112.4x + 385.3$
   (B) $y = 10.9x^2 + 385.3x - 112.4$
   (C) $y = -60.4x^2 + 7.27x - 112.4$
   (D) $y = 7.27x^2 - 60.4x + 205.44$

301

7. The number of people in a line as a function of minutes passed can be modeled with the quadratic function whose graph has a vertex of $(1, 14)$ as shown.

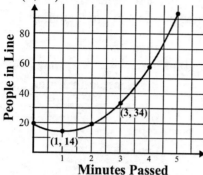

What is the quadratic function's equation?

(A) $y = 2(x - 1)^2 + 14$

(B) $y = 3(x - 1)^2 + 14$

(C) $y = 4(x - 1)^2 + 14$

(D) $y = 5(x - 1)^2 + 14$

8. Which of the following situations can best be modeled with a linear function?

(A) For every $0.15 decrease in the price of a carton of eggs, the number of cartons purchased increases by 10 percent.

(B) The total number of hot tamales sold is doubling every year.

(C) The area of a circle is equal to π times the radius squared.

(D) Every 3 months the number of potholes in the street increases by 5.

9. The change in the voltage of a car battery is decreasing at a constant rate. This can be modeled with the function $y = -x^2 + 9$. What is the domain of the model?

(A) $-9 \leq x \leq 9$

(B) $-3 \leq x \leq 3$

(C) $0 \leq x \leq 3$

(D) $0 \leq x \leq 9$

10. Heather and Robert each used eyeballing to determine the equation to model the data shown.

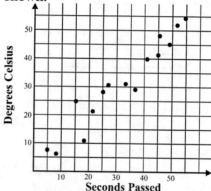

What of the following statements is true?

(A) Their equations will likely be the same because their data is the same.

(B) Their equations will likely be the same because the y-intercept of the mathematical model will obviously be 0.

(C) Their equations will likely be different because eyeballing is only an approximation.

(D) Their equations will only be different if one of them makes an error.

Practice Test 1

Part 1

You may use the formula sheet on page xii as needed.

1. Which is the graph of $f(x) = x$?

(A)

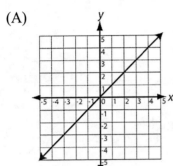

(B)

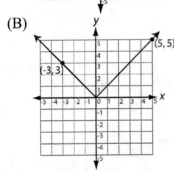

(C)

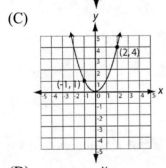

(D)

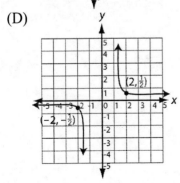

MA1A1b

2. The equation of the function $f(x) = x + 2$ changes to $y = -x - 2$. What happened to the graph of the function?

   (A) It reflects across the x-axis.
   (B) It reflects across the y-axis.
   (C) It shifts down four units.
   (D) It does not alter.

MA1A1c

3. Simplify: $f(x) = 4x - \frac{1}{2}x + \sqrt{9}$

   (A) $f(x) = 3\frac{1}{2}x + \sqrt{9}$

   (B) $f(x) = 4\frac{1}{2}x + 3$

   (C) $f(x) = 6\frac{1}{2}$

   (D) $f(x) = 3\frac{1}{2}x + 3$

MA1A2a

4. Multiply $(2x)$ and $(x^2 - 3x + 4)$. What is the answer in simplest terms?

   (A) $x^3 - 3x^2 + 4x$
   (B) $2x^3 - 6x^2 + 8x$
   (C) $3x^2 - 2x^6 + 8x$
   (D) $2x^3 + 6x^2 - 8x$

MA1A2c

5. What is $(x + 7) \times (x + 4)$ in simplest terms?

   (A) $x^2 + 11x + 28$
   (B) $x^2 + 3x + 28$
   (C) $x^2 - 3x - 28$
   (D) $x^2 + 4x + 28$

MA1A2c

6. Divide $3y^3 + 9$ by $y^9 - 9$. What is the answer in simplest terms?

(A) $\dfrac{3}{y^3 - 3}$

(B) $\dfrac{3}{y^3 + 3}$

(C) $\dfrac{3(y^3 + 3)}{y^3 + 3}$

(D) $\dfrac{3(y^3 + 3)}{y^9 - 9}$

MA1A2d

7. Factor: $t^2 - 11t + 30$

(A) $(t + 5)(t - 6)$
(B) $(t + 6)(t + 5)$
(C) $(t - 5)(t - 6)$
(D) $(t - 5)(t + 6)$

MA1A2e

8. What is the intersection point of $f(x) = 2x - 3$ and $g(x) = -x + 9$?

(A) $(4, 5)$
(B) $(1, 8)$
(C) $(6, 9)$
(D) $(12, -3)$

MA1A1i

9. Given the following image, what is x?

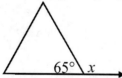

(A) $65°$
(B) $90°$
(C) $180°$
(D) $115°$

MA1G3a

10. Given the information in the triangle, what is x?

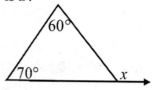

(A) $50°$
(B) $130°$
(C) $120°$
(D) $110°$

MA1G3b

11. A _____ is a generalized guess of the outcome of a situation.

(A) counterexample
(B) conjecture
(C) objecture
(D) estimate

MA1G2a

12. Given the two triangles, which theorem for triangles proves these are equivalent?

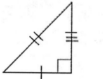

(A) SSS
(B) SAS
(C) AAS
(D) ASA

MA1G3c

13. What is the sum of the interior angles of the figure?

(A) $180°$
(B) $320°$
(C) $360°$
(D) $540°$

MA1G3a

14. What is the point of concurrency where the altitudes of a triangle intersect?

(A) Incenter
(B) Orthocenter
(C) Centroid
(D) Circumcenter

MA1G3e

15. Given the conditional statement, "If an equation is linear, then it has only one solution." What is the contrapositive of that statement?

(A) If an equation is not linear, then it does not have only one solution.
(B) If an equation has only one solution, then it is linear.
(C) If an equation does not have only one solution, then it is not linear.
(D) None of the above

MA1G2b

16. What is the inverse of the statement: "If an animal does not lay eggs, then it is a mammal?"

(A) If an animal does lay eggs, then it not is a mammal.
(B) If an animal is a mammal, then it does not lay eggs.
(C) If an animal is not a mammal, then it does lay eggs.
(D) None of the above

MA1G2b

17. If two girls are trying on shirts and there are ten different colored shirts, how many different choices can the two girls make?

(A) 5
(B) 10
(C) 20
(D) 90

MA1D1b

18. How many permutations can be made out of the letters in MATH?

(A) 4
(B) 16
(C) 20
(D) 24

MA1D1b

19. Jason has a bag of different kinds of chocolates and he picks one and then a second one without replacing the first one. What kind of event is this?

(A) Independent event
(B) Interdependent event
(C) Dependent event
(D) Not enough information

MA1D2b

20. The probability that the Red Sox and the Braves will both win a baseball game in the same day is 20%. The probability that the Braves will win is 55%. What is the probability that the Red Sox will win given that the Braves have already won?

(A) 275%
(B) 36%
(C) 64%
(D) 20%

MA1D2c

21. If you flip a coin 3 times, what is the probability you would get heads every time?

(A) 0.125
(B) 0.25
(C) 0.5
(D) 0.75

MA1D2d

22. The weights of five packages are 5 pounds, 7 pounds, 8 pounds, 5 pounds, and 10 pounds. What are the mean, median, and mode of these weights?

(A) 6.0, 5.5, 7.0
(B) 6.0, 6.5, 7.0
(C) 7.0, 7.0, 5.0
(D) 8.0, 6.0, 5.0

MA1D3a

23. What is the mean absolute deviation of $\{5, 2, 6, 1, 7, 3, 4\}$?

(A) 2

(B) $\dfrac{12}{7}$

(C) $\dfrac{7}{12}$

(D) 9

MA1D4

24. What is the mean absolute deviation of $\{3, 1, 5, 9, 4, 2, 8, 7\}$?

(A) 4.875
(B) 8
(C) 9
(D) 2.375

MA1D4

25. What are the solutions to $y = x^2 - x - 12$?

(A) $x = -4, 3$
(B) $x = 4, -3$
(C) $x = 4$
(D) $x = 3$

MA1A4b

26. In the equation $\frac{1}{2}\sqrt{x} + 1 = 3$, what is x?

(A) 1
(B) 4
(C) 9
(D) 16

MA1A2b

27. Is this function even, odd, or neither?

(A) Even
(B) Odd
(C) Neither
(D) Not enough information

MA1A1h

28. An equation's points are plotted in the table below. According to this table, how many solutions (x-intercepts) does this equation have?

x	$f(x)$
$-\frac{3}{2}$	0
$-\frac{1}{2}$	0
0	5
$\frac{1}{2}$	0
$\frac{3}{2}$	0

(A) 1
(B) 2
(C) 3
(D) 4

MA1A3c

29. In the equation $x + 5 - 2x = 6$, what is x?

(A) -1
(B) 0
(C) 1
(D) 2

30. What is the distance between $(0, 3)$ and $(-5, 6)$?

(A) 34
(B) $\sqrt{34}$
(C) 36
(D) $\sqrt{36}$

31. What is the distance between $y = -x$ and $(3, 2)$? Use the following formula:

$$d = \frac{|am + bn + c|}{\sqrt{a^2 + b^2}}$$

(A) 0
(B) 2.36
(C) -0.71
(D) 3.54

32. You are about to get a car for the first time. This means you will also get a licence plate. Each licence plate gets 3 numbers, then 3 letters. The numbers can be the digits from 0 to 9, and the letters can be any letter from A to Z. If the digits cannot be repeated and the letters cannot be repeated in the licence plate, how many possible combinations of the licence plates are there?

(A) $17,576,000$
(B) 108
(C) $11,232,000$
(D) $27,000$

33. Use the points $A = (-4, 2)$, $B = (3, 2)$, $C = (3, 5)$ and the Pythagorean Theorem to find the distance between A and C. What is the distance?

(A) 25
(B) 5
(C) 75
(D) $\sqrt{58}$

34. Graph the points $(0, 0)$, $(6, 3)$, $(3, 3)$, $(9, 0)$ to find out what shape they create. What shape do they create?

(A) a trapezoid
(B) a rectangle
(C) a rhombus
(D) a parallelogram

35. Tina has a bag of jelly beans of four flavors. There are 6 licorice, 4 grape, 3 sour apple, and 5 mint jelly beans in the bag. Tina prefers the fruit-flavored jelly beans. If she selects two jelly beans at random from the bag, what is the probability that the two jelly beans will be grape?

(A) $\dfrac{2}{51}$

(B) $\dfrac{61}{153}$

(C) $\dfrac{4}{9}$

(D) $\dfrac{4}{81}$

36. Samantha is a student at Etowah High School. Look at the following histograms. The first histogram displays the number of sodas consumed each day by each student in Samantha's math class. The second histogram displays the number of sodas consumed each day by each of the students in the ninth-grade class at her school.

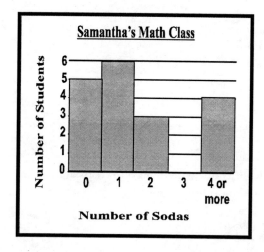

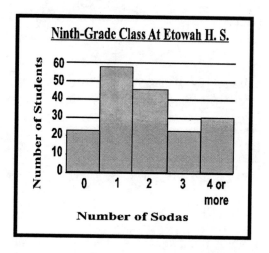

What is the approximate difference in the mode of Samantha's math class and the mode of ninth-grade class at Etowah High School?

(A) 0

(B) 1

(C) 2

(D) 3

MA1D3b

Part 2

37. Which is the graph of $f(x) = \sqrt{x}$?

(A)
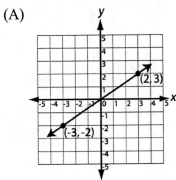

(B)
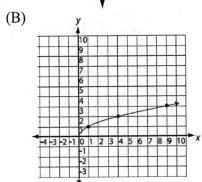

(C)
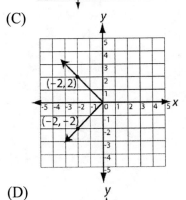

(D)
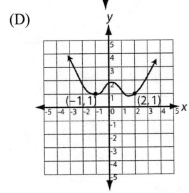

MA1A1b

38. $f(x) = \frac{1}{4}x + \frac{1}{2}$; find $f(12)$.

(A) 4.5
(B) 4
(C) 3.5
(D) 3

MA1A1a

39. The equations $y = 2$ and $y = -2$ illustrate a reflection where?

(A) over the x-axis
(B) over the y-axis
(C) around the origin
(D) There is no reflection.

MA1A1c

40. What is the minimum of $f(x) = x^2 - 1$?

(A) $(0, -1)$
(B) $(0, 0)$
(C) $(0, 1)$
(D) $(0, 2)$

MA1A1d

41. What is the pattern in the following sequence $\{1, 4, 9, 16, 25\}$?

(A) $n + 3$
(B) n^2
(C) $n^2 + 1$
(D) $n + 5(n - 2)$

MA1A1f

42. What is the rate of change of $f(x) = 3x$?

(A) 1
(B) 3
(C) $\frac{1}{3}$
(D) -3

MA1A1g

43. What is $\dfrac{\sqrt{8}}{\sqrt{2}}$ in simplest terms?

(A) $\sqrt{4}$

(B) 4

(C) $\sqrt{2}$

(D) 2

MA1A2b

44. Subtract $\dfrac{9}{x+3}$ from $\dfrac{x^2}{5}$. What is the answer in simplest terms?

(A) $x^2 - 9$

(B) $\dfrac{x^3 + 3x^2 - 45}{5x - 15}$

(C) $\dfrac{x^2 - 9}{5x + 15}$

(D) $\dfrac{x^3 + 3x^2 - 45}{5x + 15}$

MA1A2e

45. What is the volume of a cube with side length $(x + 2)$?

(A) $x^2 + 4x + 4$

(B) $x^3 + x^2 + 4x + 4$

(C) $x^3 + 6x^2 + 12x + 8$

(D) $x^3 + 12x^2 + 6x + 8$

MA1A2g

46. Solve for x: $\dfrac{3x + 2}{7} = -1$

(A) $x = 3$

(B) $x = -3$

(C) $x = -\dfrac{5}{3}$

(D) $x = \dfrac{5}{3}$

MA1A3d

47. What is the sum of the angles of a trapezoid?

(A) 180°

(B) 360°

(C) 540°

(D) 720°

MA1G3a

48. In order to form a triangle, what must the third side equal (at least) if the first two sides are 1 and 24?

(A) 3

(B) 11

(C) 24

(D) 42

MA1G3b

49. Given the two triangles, which theorem for triangles proves these are equivalent?

(A) SSS

(B) SAS

(C) AAS

(D) ASA

MA1G3c

50. Which quadrilateral has only one pair of parallel sides?

(A) Square

(B) Rhombus

(C) Parallelogram

(D) Trapezoid

MA1G3d

51. What is the name of the center of a triangle that is also the center of a circumscribed circle?

(A) Incenter

(B) Orthocenter

(C) Centroid

(D) Circumcenter

MA1G3e

52. A _____ proves conjectures to be wrong.

(A) counterexample

(B) diverse theory

(C) objecture

(D) inverse

MA1G2a

53. Given the conditional statement: p → q. The statement p → ~q is the _____.

(A) inverse

(B) converse

(C) contrapositive

(D) None of the above

MA1G2b

54. If Steven rolls two dice, what is the probability that the dice total 7 or 11?

(A) $\frac{2}{9}$

(B) $\frac{2}{12}$

(C) $\frac{1}{6}$

(D) $\frac{1}{3}$

MA1D2a

55. How many choices of ice cream sundaes are there if you can pick only one flavor out of 8 and 1 topping out of 6?

(A) 14

(B) 24

(C) 36

(D) 48

MA1D1b

56. Karen has a bag of 50 marbles, 40 are blue and ten are red. Every time she draws a blue marble, she _____ the chances of drawing a red marble.

(A) increases

(B) decreases

(C) eliminates

(D) doesn't change

MA1D2b

57. A 2003 survey revealed that 84% of homes have a garage. 65% have a garage and a backyard. What is the probability that a house has a backyard given that it has a garage?

(A) 84%

(B) 65%

(C) 80%

(D) 77%

MA1D2c

58. Daysi has a box with ten balls numbered from 0–9. What is the probability that if she picked out a numbered ball 3 times, she would get the number 7 each time?

(A) $\frac{1}{10}$

(B) $\frac{1}{100}$

(C) $\frac{3}{10}$

(D) $\frac{1}{1000}$

MA1D2d

59. Which set of numbers has the greatest range?

(A) $\{95, 86, 78, 62\}$
(B) $\{90, 65, 83, 59\}$
(C) $\{32, 29, 44, 56\}$
(D) $\{29, 35, 49, 51\}$

MA1D3a

60. In a survey of 125 randomly selected voters, 80 said they would vote for Steven Gillmor. If $20,000$ people in the district vote in the election, approximately how many would be expected to vote for Steven Gillmor?

(A) $3,125$
(B) $12,800$
(C) $14,875$
(D) $16,000$

MA1D3c

61. In the equation $10 + \sqrt{x} = 20$, what is x?

(A) 100
(B) 50
(C) 10
(D) 5

MA1A3b

62. According to the graph, what are the solution(s) (x-intercepts) to the equation?

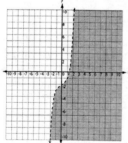

(A) $x = -2, 2$
(B) $x = -3, 3$
(C) $x = -1, 1$
(D) $x = 0$

MA1A3c

63. What is the distance between $(-1, -1)$ and $(0, 5)$?

(A) 6
(B) $\sqrt{36}$
(C) 7
(D) $\sqrt{37}$

MA1G1a

64. What is the shortest distance between $y = 4x + 1$ and $(2, 3)$? Use the following formula:
$$d = \frac{|am + bn + c|}{\sqrt{a^2 + b^2}}$$

(A) 1.46
(B) -1.46
(C) 2.91
(D) Cannot be determined

MA1G1b

65. What is the midpoint of $(6.4, 3)$ and $(-10.7, 4)$?

(A) $(3, 2)$
(B) $(-2.15, 3.5)$
(C) $(-2, 3)$
(D) $(2, -3)$

MA1G1c

66. What is the midpoint of $(-3, 4)$ and $(2, 1)$?

(A) $\left(-\frac{1}{2}, 2\frac{1}{2}\right)$
(B) $(-1, 2)$
(C) $(1, -2)$
(D) $(1, 2)$

MA1G1c

67. Use the points $A = (-4, -4)$, $B = (-4, 2)$, $C = (3, 2)$ and the Pythagorean Theorem to find the distance between A and C. What is the distance?

(A) 100
(B) $\sqrt{85}$
(C) 7
(D) $\sqrt{110}$

MA1G1d

68. What is the interquartile range of the data below?

$$\{1, 1, 2, 3, 7, 8, 10\}$$

(A) 9
(B) 6
(C) 7
(D) 1

MA1D3a

69. Graph the points $(0, 0)$, $(1, 2)$, $(4, 0)$, $(5, 2)$ to find out what shape they create. What shape do they create?

(A) a square
(B) a rectangle
(C) a rhombus
(D) a parallelogram

MA1G1e

70. Look at the two figures below. The figure on the left represents the after school activities of one Georgia high school. The figure on the right represents the after school activities for ALL Georgia high schools. What is the ratio of the children who play sports from one high school to all high schools?

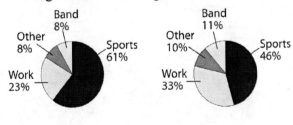

(A) 60 : 40
(B) 61 : 46
(C) 3 : 2
(D) 39 : 54

MA1D3b

71. You are about to get a car for the first time. This means you will also get a licence plate. Each licence plate gets 3 numbers, then 3 letters. The numbers can be the digits from 0 to 9, and the letters can be any letter from A to Z. There are no restrictions on what numbers you can get, but you cannot have any repeating letters. Based on this information, how many possible combinations of the licence plates are there?

(A) 11, 232, 000
(B) 15, 600, 000
(C) 17, 576, 000
(D) 12, 654, 720

MA1D1a

72. Jon and Nick took 5 tests in their math class this semester. They recorded their scores for each test in the chart below.

Test	Jon	Nick
1	92	84
2	85	89
3	90	96
4	78	84
5	95	92

Who had the greater mean deviation for his score?

(A) Jon
(B) Nick
(C) Neither, they had the same mean deviation.
(D) This is not possible given the information provided.

MA1D4

Practice Test 2

Part 1

You may use the formula sheet on page xii as needed.

1. $f(x) = \frac{1}{2}x + 7$; find $f(8)$.

   (A) 15

   (B) 4

   (C) 11

   (D) 18

   MA1A1a

2. According to the graph, as the x value increases, the y value _____.

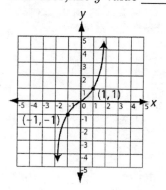

   (A) increases

   (B) decreases

   (C) gets shorter

   (D) gets wider

   MA1A1e

3. What is $x + 2 + \sqrt{4}$ in simplest terms?

   (A) $2x + \sqrt{4}$

   (B) $x + 2\sqrt{4}$

   (C) $x + 4$

   (D) $x + 2 + 2$

   MA1A2a

4. What is $bc^2 - 8bc - 2b^2c^2$ divided by $2bc$?

   (A) $\frac{c}{2} - 4 - bc$

   (B) $-\frac{c}{2} + 4 + bc$

   (C) $-\frac{2}{c} + 4 + \frac{1}{bc}$

   (D) $\frac{2}{c} - 4 - \frac{1}{bc}$

   MA1A2c

5. After multiplication, what is $(x - 1)(x - 3)$ in simplest terms?

   (A) $x^2 - 4x + 3$

   (B) $x^2 - 4x - 3$

   (C) $x^2 + 3x - 4$

   (D) $x^2 - 3x + 4$

   MA1A2e

6. What is $64c^2 - 25$ divided by $8c^2 - 5$?

   (A) $(8c - 5)(8c + 5)$

   (B) $\dfrac{(8c - 5)(8c + 5)}{8c - 5}$

   (C) $8c + 5$

   (D) $\dfrac{(8c - 5)(8c + 5)}{8c^2 - 5}$

   MA1A2d

7. Using greatest common factors, what are the factors of $20y^3 + 30y^5$?

(A) $5y^3(4 + 6y^2)$
(B) $10y^3(2 + 3y^2)$
(C) $10y^3(4 + 6y^2)$
(D) $10y^5(2 + 3y^2)$

MA1A2e

8. What is the value of x in the equation $-\sqrt{x} + 10 = 7$?

(A) $x = 289$
(B) $x = -289$
(C) $x = 9$
(D) $x = -9$

MA1A2b

9. What is the sum of the interior angles of a square?

(A) $90°$
(B) $180°$
(C) $360°$
(D) $540°$

MA1G3a

10. Using the diagram, which angle has the largest measure?

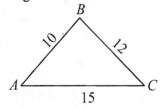

(A) $\angle A$
(B) $\angle B$
(C) $\angle C$
(D) Not enough information

MA1G3b

11. How many different ways can you arrange 5 different pictures on a bookshelf?

(A) 5
(B) 20
(C) 60
(D) 120

MA1D1b

12. Given the two triangles, which theorem for triangles proves these are equivalent?

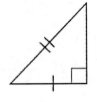

(A) SSS
(B) SAS
(C) HL
(D) SSA

MA1G3c

13. Which quadrilateral does the figure represent?

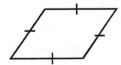

(A) Rectangle
(B) Square
(C) Rhombus
(D) Triangle

MA1G3d

14. In a right triangle, which point of concurrency lies on the 90° angle?

(A) incenter
(B) centroid
(C) orthocenter
(D) circumcenter

MA1G3e

15. If I said 60 is divisible by all numbers less than ten, to prove me wrong, you would give a(n) _____.

(A) inverse

(B) counterexample

(C) conjecture

(D) converse

MA1G2a

16. Given the statement, "If two triangles are congruent, then they are similar." What is the inverse of that statement?

(A) If two triangles are not congruent, then they are not similar.

(B) If two triangles are similar, then they are congruent.

(C) If two triangles are not similar, then they are not congruent.

(D) None of the above

MA1G2b

17. Given the conditional statement p → q, what does q → p represent?

(A) Inverse

(B) Converse

(C) Contrapositive

(D) None of the above

MA1G2b

18. Which equation represents the number of one thing OR another?

(A) $p(A \text{ or } B) = p(A) + p(B) - p(A \cap B)$

(B) $p(A \text{ or } B) = p(A) + p(B) + p(A \cap B)$

(C) $p(A \text{ or } B) = p(A) + p(B) - p(A \cup B)$

(D) $p(A \text{ or } B) = p(A) + p(B) + p(A \cup B)$

MA1D1a

19. Which equation represents the combination formula?

(A) $_nC_r = \dfrac{n}{n-r}$

(B) $_nC_r = \dfrac{n}{(n-r)r}$

(C) $_nC_r = \dfrac{n!}{(n-r)!r!}$

(D) $_nC_r = \dfrac{n!}{(n-r)!}$

MA1D1b

20. Cindy has a bag of 20 marbles. There are 10 red marbles, 7 black marbles, and 3 blue marbles in the bag. What is the probability that Cindy will draw a red, then a black marble (without replacement)?

(A) $\frac{17}{20}$

(B) $\frac{7}{38}$

(C) $\frac{7}{40}$

(D) $\frac{17}{40}$

MA1D2b

21. A pair of dice is rolled. What is the probability that the sum of the two dice will be greater than four given that the first die rolled is one?

(A) 0.083

(B) 0.167

(C) 0.5

(D) 0.33

MA1D2c

22. What is the mean absolute deviation of {9, 4, 8, 3, 6}?

(A) 1

(B) 1.5

(C) 2

(D) 2.5

MA1D4

23. A box of a dozen donuts has 3 lemon cream-filled, 5 chocolate cream-filled, and 4 vanilla cream-filled. If the donuts look identical, what is the probability of picking a lemon cream-filled?

(A) $\frac{3}{10}$

(B) $\frac{3}{5}$

(C) $\frac{3}{11}$

(D) $\frac{1}{4}$

MA1D2d

24. Rachel kept track of how many scoops she sold of the five most popular flavors in her ice cream shop.

Flavor	Scoops
Vanilla Bean	30
Chunky Chocolate	36
Strawberry Coconut	44
Chocolate Peanut Butter	46
Mint Chocolate Chip	??

How many scoops of mint chocolate chip would need to be sold to make the mean number of the scoops sold 38?

(A) 34

(B) 35

(C) 36

(D) 38

MA1D3a

25. What is the mean absolute deviation of $\{7, 3, 7, 2, 5, 4, 8, 1\}$?

(A) 2

(B) 2.125

(C) 3

(D) 3.125

MA1D4

26. What are the solutions to $y = x^2 + 5x - 6$?

(A) $x = 6, 1$

(B) $x = -6, -1$

(C) $x = -6, 1$

(D) $x = 6, -1$

MA1A4b

27. Is the equation represented by this graph even, odd, or neither?

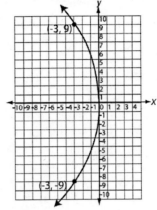

(A) Even

(B) Odd

(C) Neither

(D) Not enough information

MA1A1h

28. Given the graph, what is the solution (x-intercept) to the equation of the graph?

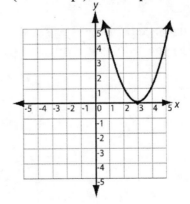

(A) $x = 3$

(B) $x = 0$

(C) $x = 2$

(D) $x = 2.8$

MA1A3c

29. Using the intersection points of the graphs of the equations, what are the solutions of $f(x) = x^2 - 7x - 8$ and $g(x) = 7x + 7$?

(A) $x = 15, -1$
(B) $x = -15, 1$
(C) $x = 7, 8$
(D) $x = -8, 7$

MA1A1i

30. In the equation $2x - 3 = x + 4$, what is x?

(A) $\dfrac{2}{3}$

(B) $-\dfrac{2}{3}$

(C) 7

(D) $\dfrac{3}{4}$

MM1A3d

31. What is the distance between $(-1, 1)$ and $(3, -4)$?

(A) 41
(B) $\sqrt{41}$
(C) 40
(D) $\sqrt{40}$

MA1G1a

32. What is the distance between $y = \frac{1}{2}x + 6$ and $(3, 0)$? Use the following formula:

$$d = \frac{|am + bn + c|}{\sqrt{a^2 + b^2}}$$

(A) $\dfrac{1}{2}$

(B) $\dfrac{1}{4}$

(C) 6.71
(D) 12

MA1G1b

33. Use the points $A = (-4, -1)$, $B = (-4, -4)$, $C = (-1, -1)$ and the Pythagorean Theorem to find the distance between C and B. What is the distance?

(A) $\sqrt{18}$

(B) 18

(C) $\sqrt{81}$

(D) 81

MA1G1d

34. What kind of triangle do the points $(-3, 0)$, $(3, 0)$, $(0, 3)$ make?

(A) Scalene

(B) Equilateral

(C) Isosceles

(D) They don't form a triangle.

MA1G1e

35. You have a standard deck of 52 cards that has 4 each of 13 kinds of cards. If you draw one card from the deck, what is the probability that the card will be either a Jack or a nine?

(A) $\dfrac{2}{13}$

(B) $\dfrac{1}{13}$

(C) $\dfrac{1}{26}$

(D) $\dfrac{3}{26}$

MA1D2a

36. Samantha is a student at Etowah High School. Look at the following histograms. The first histogram displays the number of sodas consumed each day by each student in Samantha's math class. The second histogram displays the number of sodas consumed each day by each of the students in the ninth-grade class at her school.

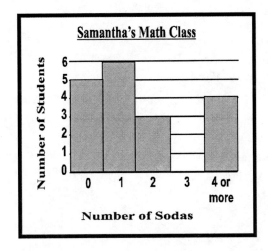

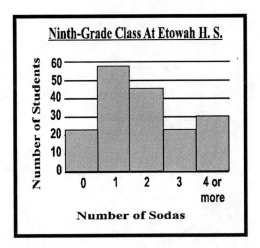

What is the difference in the median of Samantha's math class and the median of ninth-grade class at Etowah High School?

(A) 0

(B) 1

(C) 2

(D) 3

MA1D3b

Part 2

37. Which is the graph of $f(x) = |x|$?

(A)

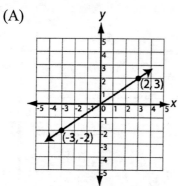

(B)

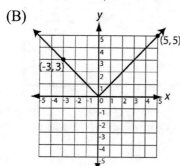

(C)

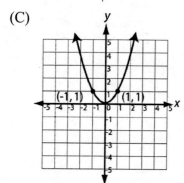

(D)

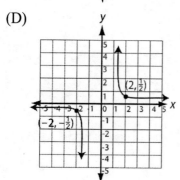

MA1A1b

38. The graph of $y = 6x + 1$ is reflected over the y-axis. What is the new equation of the graph?

(A) $y = -6x + 1$
(B) $y = -6x - 1$
(C) $y = 6x - 1$
(D) $y = \frac{1}{6}x + 1$

MA1A1c

39. What is the pattern in the following sequence $\{-2, -4, -6, -8\}$?

(A) $-2n$
(B) $2n$
(C) $-\frac{1}{2}n$
(D) $2n + 1$

MA1A1f

40. In the equation $f(x) = x^2 + x + 1$, the rate of change is _____.

(A) constant
(B) variable
(C) nonexistent
(D) not enough information

MA1A1g

41. What is $\dfrac{\sqrt{2}}{\sqrt{6}}$ in simplest terms?

(A) $\dfrac{\sqrt{2} \times \sqrt{6}}{6}$

(B) $\dfrac{\sqrt{12}}{6}$

(C) $\dfrac{2\sqrt{3}}{6}$

(D) $\dfrac{\sqrt{3}}{3}$

MA1A2b

42. After multiplying, what is $2x(x^2 + 3x - 7)$ in simplest terms?

(A) $2x^3 + 6x^2 - 14x$
(B) $2x^3 + 6x^2 + 14x$
(C) $2x^2 + 6x - 14$
(D) $2x^3 + 3x^2 - 14x$

MA1A2c

43. What is the sum of $\dfrac{5}{y^2}$ and $\dfrac{6}{y^2}$?

(A) $\dfrac{11}{2y^2}$

(B) $\dfrac{5+6}{y^2}$

(C) $\dfrac{11}{y^2}$

(D) $\dfrac{5}{y^2} + \dfrac{6}{y^2}$

MA1A2d

44. What is the volume of a rectangular prism with length $(x + 1)$, width $(2x - 3)$, and height (x^2)?

(A) $2x^3 - x - 3$
(B) $2x^4 - x^3 - 3x^2$
(C) $2x^4 + x^3 + 3x^2$
(D) $2x^4 + x^3 - 3x^2$

MA1A2f

45. Using the diagram, what does E equal?

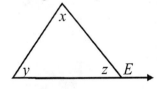

(A) $x + y$
(B) $y + z$
(C) $z + x$
(D) $x + y + z$

MA1G3b

46. Given the two triangles, which theorem for triangles proves these are equivalent?

(A) SSS
(B) AAS
(C) ASA
(D) SAS

MA1G3c

47. Which quadrilateral does the figure represent?

(A) Rectangle
(B) Square
(C) Rhombus
(D) Triangle

MA1G3d

48. The _____ bisects the medians.

(A) incenter
(B) centroid
(C) orthocenter
(D) circumcenter

MA1G3e

49. A(n) _____ negates the hypothesis and the conclusion of a statement.

(A) conjecture
(B) counterexample
(C) converse
(D) inverse

MA1G2a

50. Given the conditional statement: p → q, what does ∼p → ∼q represent?

(A) Inverse
(B) Converse
(C) Contrapositive
(D) None of the above

MA1G2b

51. If Erin picks up a piece of ice and notices that is cold, then says, "All ice is cold," this is an example of

(A) deductive reasoning.
(B) inductive reasoning.
(C) deduction.
(D) conjection.

MA1G2a

52. When Katie walked into the ice cream shop, she counted the number of flavors (3) and the number of toppings (8). Assuming she can only get one topping and one flavor, how many choices of combinations does Katie have?

(A) 11
(B) 16
(C) 24
(D) 32

MA1D1b

53. What is the formula for mutually exclusive events?

(A) $n(A \cup B) = n(A) + n(B) - n(A \cap B)$

(B) $\dfrac{n!}{(n-r)!r!}$

(C) $\dfrac{2pk}{4r}$

(D) $P(A \text{ or } B) = P(A) + P(B)$

MA1D2a

54. The weights of 8 puppies are 10 pounds, 10 pounds, 11 pounds, 8 pounds, 11 pounds, 13 pounds, 6 pounds, and 11 pounds. What are the mean, median, and mode of these weights?

(A) 10, 10, 10
(B) 10, 10.5, 11
(C) 10, 11, 11
(D) 10, 10, 11

MA1D3a

55. Blake has a bag of 35 marbles. There are 10 white marbles, 15 yellow marbles, and 10 blue marbles in the bag. What is the probability that Blake will draw a white, then a yellow marble (without replacement)?

(A) $\dfrac{35}{69}$

(B) $\dfrac{15}{69}$

(C) $\dfrac{30}{119}$

(D) $\dfrac{15}{119}$

MA1D2b

56. A pair of dice is rolled. What is the probability that the sum of the dice is less than 10, given that the first die is four?

(A) $\dfrac{5}{36}$

(B) $\dfrac{5}{6}$

(C) $\dfrac{1}{6}$

(D) 1

MA1D2c

57. If you have a normal set of dice, what is the probability that if you roll it 3 times, you will get the number 2 each time?

(A) $\dfrac{1}{6}$

(B) $\dfrac{1}{36}$

(C) $\dfrac{1}{216}$

(D) $\dfrac{1}{360}$

MA1D2d

58. Tom's school was considering making uniforms mandatory starting with the next school year. Tom hated the idea and wanted to do his own survey to see if parents were really in favor of it. He considered 4 places to conduct his survey. Which would give the most valid results?

(A) He would stop people at random walking through the mall.
(B) He would survey parents in the car pool lanes picking up students after school.
(C) He would survey the teachers after school.
(D) He would survey the students in his biology class to ask what their parents thought.

MA1D3c

59. What are the solutions to $y = x^2 - 2x - 3$?

(A) $x = 3, 1$
(B) $x = -3, 1$
(C) $x = -3, -1$
(D) $x = 3, -1$

MA1A4b

60. In the equation $3\sqrt{x} - 1 = 17$, what is x?

(A) 6
(B) 36
(C) 18
(D) 46

MA1A3b

61. The points of the graph of an equation are plotted in the table below. Given this table, what are the solution(s) (x-intercepts) to the equation?

x	0	1	2	3	4	5	6
$f(x)$	16	9	4	1	0	1	4

(A) $x = 4$
(B) $x = 16$
(C) $x = 6, 2$
(D) $x = 3, 5$

MA1A3c

62. In the equation $x^2 + 4x = 7x + 18$, what is x?

(A) $x = 6, 3$
(B) $x = 9, -2$
(C) $x = -6, 3$
(D) $x = 6, -3$

MM1A3d

63. What is the distance between $(1, 7)$ and $(-1, -1)$?

(A) $\sqrt{69}$
(B) $\sqrt{68}$
(C) 8
(D) $\sqrt{67}$

MA1G1a

64. What is the distance between $y = -4x + 7$ and $(4, 4)$? Use the following formula:

$$d = \frac{|am + bn + c|}{\sqrt{a^2 + b^2}}$$

(A) 3.15
(B) -3.15
(C) 6.55
(D) -6.55

MA1G1b

65. What is the midpoint of $(5, 4)$ and $(-3, 4)$?

(A) $(0, 4)$

(B) $(-1, 4)$

(C) $(4, 1)$

(D) $(1, 4)$

MA1G1c

66. Use the points $A = (1, -1)$, $B = (3, -1)$, $C = (3, 2)$ and the Pythagorean Theorem to find the distance between A and C. What is the distance?

(A) 9

(B) 4

(C) 13

(D) $\sqrt{13}$

MA1G1d

67. Use the points $A = (3, 2)$, $B = (3, -1)$, $C = (-4, -1)$ and the Pythagorean Theorem to find the distance between A and C. What is the distance?

(A) 3

(B) 9

(C) $\sqrt{10}$

(D) $\sqrt{58}$

MA1G1d

68. What shape do the points $(-4, 0)$, $(-4, 2)$, $(4, 0)$, $(4, 2)$ make?

(A) Rectangle

(B) Square

(C) Parallelogram

(D) Rhombus

MA1G1e

69. The figure on the left depicts sports played by seniors in one high school. The figure on the right depicts sports played by seniors in 100 different high schools. Would figure 1 be a decent representation of all high schools?

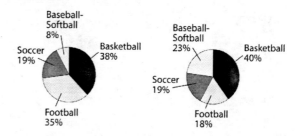

(A) Yes, the proportion of varying sports is about the same.

(B) Yes, because there are a lot of seniors at one high school.

(C) No, the proportions of the sports are not similar.

(D) No, this doesn't provide us with a properly diverse sample.

MA1D3b

70. Cynthia and Robert work at the local video store in their town. The table below shows the amount of hours they worked each week during the month of August.

Week	Cynthia's Hours	Robert's Hours
Week 1	23	29
Week 2	35	31
Week 3	25	30
Week 4	32	28

What is the difference in the median of Cynthia's hours and the median of Robert's hours?

(A) 0.75

(B) 1

(C) 3.25

(D) 9

MA1D3a

71. Nicole is in charge of buying food for a political campaign picnic lunch in City Center Park. In order to determine which foods people would prefer, Nicole conducted a survey of food preferences in front of a small health-food store near the park. The results indicated that most people prefer fruits and vegetables.

Which of the following would be one way that you could improve upon the accuracy of this survey?

(A) She could improve her survey by taking a sample of what types of food her family would prefer.
(B) She could improve her survey by taking a random sample for the population of people that do not like politics.
(C) She could not improve her survey because she picked the best way to survey people.
(D) She could improve her survey by taking a random sample for the population of people who might come to the lunch.

MA1D3c

72. Use number set A to answer the question.

$$A = \{-11, -7, -3, -1, 0, 2, 4, 8, 10, 16\}$$

How many ways can you choose a negative or even number from A?

(A) 11
(B) 10
(C) 9
(D) 7

MA1D1a

Index